AF479955

Space Science & Technology for Geographical Research & Applications

Space Science & Technology for Geographical Research & Applications

Editors

N. C. Gautam

V. Raghavswamy

BSP **BS Publications**

4-4-309, Giriraj Lane, Sultan Bazar, Hyderabad - 500 095. (A.P.)

Ph : 23445600, 23445688. Fax : 91 + 40 - 23445611

e-mail: contactus@bspublications.net

Published by :

BS Publications

4-4-309, Giriraj Lane, Sultan Bazar,
Hyderabad - 500 095 A.P.
Phone : 040 - 23445601, 23445688, 23445600
e-mail : contactus@bspublications.net

ISBN : 978-93-90211-12-8

Foreword

Geography is very well reckoned as mother of all sciences. It is well understood, read and taught all over the world. Over the years the subject of Geography has graduated from a mere description of the various geographical facets of the Earth to qualitative description of the spatial distribution of features to quantitative estimation in the context of use of digital methods and techniques. Space Science, Space Technology and its applications have contributed significantly to understand and capture the geography of Earth, its natural processes, the man made changes and their positive and negative impacts on the Earths system. With the newer developments in space technologies and its applications in the future, I am sure this will provide further insight in understanding the mother Earth.

I congratulate the Centre for Land Use Management (CLUMA), Hyderabad for taking the initiative in organising such an important National Conference on "Space Science and Technology for Geographical Research and Applications" in association with NRSA, ISRO/DOS, JNTU and Bhoovigyan Foundation, New Delhi. I am sure the papers published in the proceedings will enrich the knowledge of the geographic community of teachers, researchers and students. I congratulate Dr. N.C. Gautam and Dr. V. Raghavswamy in bringing out this important publication.

Hyderabad
September 2005

Prof. B.L. Deekshatulu
Former Director,
NRSA & UN-CSSTE-AP and
Visiting Professor
HCU, Hyderabad

Preface

Space science and technology has grown multifold globally and Indian contribution to its growth over the years has been very significant. Further, its contribution in the domain of use of space derived data to a variety of Applications has also grown multifold, ever since the launch of IRS series of satellites in 1988. The wide research of Applications stands spread horizontally across the geographical boundaries, regions/sub-regions and cultural landscapes. This capability has also grown from vertical integration at different spatial scales, time series and platforms (satellite, aerial and ground) with the availability of GIS and Geoinformatics techniques. In the recent years, convergent of technologies like geospatial or geoinformatics have opened new vistas and challenges in the area of Earth Observation (EO) studies. Here, one could come across a number of good examples as amply demonstrated in the variety of collection of research articles as presented in this compendium of proceedings. Earth Observation (EO) from remote sensing platforms offer synoptic view, repetitive coverage of multispectral and spatial resolutions data, over macro, meso and micro areas. Such a data is highly useful to collate, inventorise, evaluate and analyse the geosphere and biosphere resources and the processes with a high degree of reliability, accuracy and in a cost effective manner. In addition, GPS/GIS allows to integrate and model spatial and non-spatial data to oversee planning, management and decision making process. In the recent past, the telecommunication space systems are also strengthening the earth observation capability in the areas like tele-disaster, tele-medicine and tele-education in the country.

Geography plays a key role in understanding a real differentiation and inter-relationships between *'man activities and environmental elements'* in the *'area-region-time scale'* which would help to provide solutions to minimize the intra and inter-regional disparities to achieve a balanced economic and social development. Geography, traditionally has been a multi-disciplinary *(cartography – demography – regional planning)* discipline. It is heavily dependent upon the 'data' (primary and socio-economic) to apply geographical theories for appropriate geographical solutions in the area of spatial planning and regional development. It has immensely been benefited from remote sensing and GIS, which offer a wide variety of data and analytical techniques.

Recognising the importance of the theme of the conference, the contributions of papers have been broadly grouped into the following sub-themes : i) Space science and space technology developments in India ii) Earth Observations (EO) and its applications scenario iii) Geographical Research : Issues and solutions   iv) Natural Resources and Repository : soil, ground water, wetlands and land use   v) Watershed management   vi) GIS technology and applications vii) Urban and regional studies and   viii) Disaster management.

We hope the discussions emerging out of the above sub themes during the conference will further strength and offer newer initiatives in undertaking research and applications in geography.

It is our pleasure to record our sincere thanks to Sri. G. Madhavan Nair, Chairman, ISRO and Secretary, DOS, Govt. of India, Bangalore for his guidance and encouragement and to Dr. R.R.Navalgund, Director, NRSA/DOS, Prof. K.Rajagopal, Vice Chancellor, JNTU, Prof. Y.Anjaneyulu, Director, Institute of Science & Technology (IST), JNTU, Dr. K.V.Sundaram, Chairman, Bhoovigyan Vikas Foundation, New Delhi Dr. P.S.Roy, Deputy Director, RS&GIS-Applications Area, NRSA and to Dr. I.V.Muralikrishna, Head, Centre for Spatial Information Technology, JNTU for their wholehearted support.

We express our thanks to all the authors who have submitted their valuable contribution of papers, sharing their knowledge, experience and wisdom. The various suggestions and guidance extended by the organising committee of the Conference is also gratefully acknowledged.

Finally, we thank the B.S.Publications, Hyderabad for publishing the proceedings.

Hyderabad

September 2005

N. C. Gautam
V. Raghavswamy

Contents

Space Science &
Space Technology Development in India

Satellite Data Products and Services

Jospeh Arokiadas and K.P.R.Menon
NDC, NRSA, Hyderabad

ABSTRACT

The National Remote Sensing Agency (NRSA) of Department of Space (DOS), Govt of India is the focal point for the distribution of satellite data products in India and neighboring countries. NRSA has an earth station at Shadnagar about 55 KM from Hyderabad to receive the data from all contemporary remote sensing satellites such as CARTOSAT, RESOURECSAT, IRSP4, IRS1D, IRS1C etc and other foreign mission like NOAA series, TERRA etc. The paper gives an over view the entire Indian Remote Sensing (IRS) mission and their sensor specifications. The foreign mission data, which are distributed, and their details are also briefed. The capabilities of different mission to target specific sites in different spectral regions are briefed. The stereo viewing and flexibility in viewing for better revisit and programming is discussed. The product generation is through an automated environment and in order to avoid ambiguities in the generation of data a nine-digit product codes are defined. The details of these codes and types of products are explained in this paper.

Introduction

The National Remote Sensing Agency (NRSA), of the Dept. of Space (DOS), Govt. of India (GOI), is the focal point for distribution of remote sensing satellite data products in India and its neighboring countries. NRSA has an earth station at Shadnagar, about 55 km from Hyderabad, to receive data from almost all contemporary remote sensing satellites such as CARTOSAT (IRS-P5), RESOURCESAT (IRS-P6), IRS-P4, IRS-1D, IRS-1C, IRS-P3, IRS-1B, IRS1A, ERS-1/2, NOAA, AQUA, TERRA series of satellites. Mission wise details for regarding sensors and products are provided in the following section. The paper focuses on the IRS missions and the related services from NDC.

The data is recorded at Shadnagar on High Density Digital Tapes (HDTs) or Digital Linear Tapes (DLTs) or CD-ROMs or 8 mm Exabyte Digital Tapes (DATs) depending on the mission, and archived for providing data products to users as and when orders are received. As per the archival policy data will be archived for a period of 5 years before being selectively purged.

Data products can be supplied on a wide variety of media and formats such as:

❖ **Photographic products**

♦ B&W and False Colour Composites (FCC)
♦ Positive transparencies (240 mm)
♦ Negative transparencies (240 mm)
♦ Paper prints (1x, 2x, 4x and 5x enlargements up to 1000 mm width)

❖ **Digital products**

♦ CD-ROMS
♦ 8 mm Exabyte Tapes (DATs)

NRSA Data Centre, which is the focal point for satellite data dissemination, assists users in selecting the appropriate data based on the application and availability of cloud free data.

IRS Satellite Missions

NRSA has been acquiring satellite data from the late seventies and has an excellent archival of historical and current data. The lists of satellite missions along with the payload characteristics (such as resolution, swath, spectral bands), revisit, etc., are given in Annexure-I. It can be seen that users have a wide choice to meet their requirements. The Indian Remote sensing Satellites are an important element of National Natural Resource Management System (NNRMS) and serve a national goal in terms of providing continuous and operational remote sensing data services for effective management of India's natural resources.

IRS-1A/1B

The Indian Remote Sensing Satellite (IRS) system became operational with the launch of IRS-1A, on March 17,1988. The second satellite, IRS-1B, identical to IRS-1A, was launched on August 29,1991 and is still operational.

These were operational, first generation remote sensing satellites with two Linear Imaging Self Scanning Sensors (LISS-1 & LISS-II) for providing data in four spectral bands (in visible and near infra-red regions) with a spatial resolution of 72.5m and 36.25m, with repetivity of 22 days.

IRS-P2 / IRS-P3

IRS-P2 and IRS-P3 launched by India's PSLV on October 15,1994 and March 21,1996 respectively are the first two satellites in the IRS-P series intended for technology proving missions. The IRS-P2 is similar to IRS-1A and 1B and carried a modified LISS-II camera.

IRS-P3 carries a WiFS camera similar to that of IRS-1C but with an additional spectral band SWIR. Besides it carries a Modular Opto-Electronic Scanner (MOS) of the German

Space Agency, DLR and an X-ray astronomy payload. MOS is useful for developing algorithms for ocean applications.

IRS-1C/1D

With the successful design, fabrication, development, launch, in-orbit performance of the first generation of IRS, India surged ahead to provide improved and enhanced data services with the second generation remote sensing satellites, viz., IRS-1C/1D.

IRS-1C was launched on December 28,1995 and IRS-1D was launched on Sept. 29, 1997. Both the satellites are characterized by improved spatial resolution, extended spectral bands, stereo viewing and faster revisit capability. IRS-1C/1D has three cameras onboard namely: Panchromatic camera (PAN), which is a high resolution camera operating in panchromatic band with a resolution of 5.8m and swath of 70 kms. PAN camera can be steered up to + 26° across the track to provide stereoscopic data and to improve the revisit to 5 days. The Linear Imaging Self-scanning Sensor (LISS-III) operating in four bands – two in Visible, one in Near Infrared (VNIR) and one in Short wave Infrared (SWIR) ranges. It provides a ground resolution of 23.5 m in VNIR bands and 70.5 m in SWIR band with a swath of 141 kms and 148 kms in V/NIR and SWIR bands respectively. LISS-III data provides continuity to users who are familiar with the usage of IRS-1A/1B LISS-II and TM data. The Wide Field Sensor (WiFS) a coarse resolution camera with spatial resolution of 188 m operates in two bands with a swath of 810 kms. Due to its large swath, WiFS provides a revisit of 5 days.

IRS-P4 (Oceansat)

The IRS-P4 satellite, which is also called as Oceansat, primarily caters to oceanographic applications. The indigenous Polar Satellite Launch Vehicle (PSLV) launched this satellite on 26th May 1999, into a polar Sun-synchronous orbit at an altitude of 720 km. The satellite has a high repetivity of 2 days. The payload includes an Ocean Colour Monitor (OCM), a Multifrequency Scanning Microwave Radiometer (MSMR) and solid-state memory for recording data outside the visibility of a ground station.

The OCM has 8 spectral bands in the visible and near infrared region of the electromagnetic spectrum. The data collected by this sensor is optimum for quantitative estimation of ocean primary productivity. The resolution of the OCM is 360mx250m. With this resolution, it is expected to get better information on chlorophyll distribution near the coast. High radiometric sensitivity and dynamic range are provided to measure the reflectance, varying from 0.7% to 7%, from the ocean surface. Four bands 0.545-0.565, 0.660-0.680,0.745-0.785 and 0.845-0.885 microns have a dynamic range covering 100% solar reflectance, thus making it suitable for land applications including cloud/snow studies. The MSMR operates in 4 frequencies with two polarizations. Many geophysical parameters such as Sea Surface Temperature (SST), wind speed over oceans, total precipitable water in the atmosphere etc., strongly influence the black body radiation from earth's surface. It is therefore possible to estimate a number of such parameters by passive microwave radiometers. Microwave measurements have the additional advantage of all weather capability.

IRS-P6 (RESOURCESAT)

The IRS-P6 is envisaged as the continuity mission to IRS-1C/1D, with enhanced capabilities both in the payload and the platform, to meet the increasing demands of the user community. The indigenously built Polar Satellite Launch Vehicle launched the satellite on October 17, 2003. The orbit parameters of IRS-P6 are same as IRS-1C. The payload system of IRS-P6 consists of three solid state cameras A high resolution multispectral sensor - LISS-IV, A medium resolution multispectral sensor - LISS-III and coarse resolution Advanced Wide Field Sensor - AWiFS

The LISS-IV camera is a multispectral high-resolution camera with a spatial resolution of 5.8m at nadir. The sensor consists of three linear odd-even pairs of CCD arrays, each with 12000 pixels. This camera can be operated in two modes: MONO and multi-spectral. In the multispectral mode, data is collected in three spectral bands – Green, Red and NIR.In the multispectral mode, the sensor provides data corresponding to pre-selected 4096 contiguous pixels, corresponding to 23.9 Km swath. The 4K strip can be selected anywhere within the 12K pixels by commanding the start pixel number using electronic scanning scheme. In MONO mode, the data of full 12K pixels of any one selected band, corresponding to a swath of 70 Km, can be transmitted. Nominally, Red band data are transmitted in this mode. The LISS-IV camera has the additional feature of off-nadir viewing capability by tilting the camera by + 26°. This way it can provide a revisit of 5 days for any given ground area.

The LISS-III is a multi-spectral camera operating in four spectral bands, three in the visible and near infrared and one in SWIR region, as in the case of IRS-1C/1D. The new feature in LISS-III camera is the SWIR band (1.55 to 1.7 microns), which provides data with a spatial resolution of 23.5m unlike IRS-1C/1D (the spatial resolution is 70 Km).

The AWiFS camera provides enhanced capabilities compared to the WiFS camera on-board IRS-1C/1D, in terms of spatial resolution (56 m Vs 188m), radiometric resolution (10 bits Vs 7 bits) and Spectral bands (4 Vs 2) with the additional feature of on-board detector calibration using LEDs. The spectral bands of AWiFS are same as LISS-III. The AwiFS has a wide swath of 740 KM.

In addition to the three sensors, the satellite carries an OBSSR for collecting data outside the vicinity of the ground stations. The OBSSR has a capacity of 120 GB. It can record data of 36 minutes: LISS-III, AWiFS and LISS-IV for 4.5 minutes, only LISS-IV or LISS-III + AWiFS data for 9 minutes.

IRS-P5 (CARTOSAT)

The IRS-P5 or CARTOSAT-1 is envisaged as a mission to meet the stereo data requirements of the user community. CARTOSAT-1 is a global mission. The nominal life of the mission is planned to be five years. The indigenously built Polar Satellite Launch Vehicle launched the satellite on May 05, 2005.

The payload system of IRS-P5 consists of two Panchromatic solid state cameras - Fore and Aft, mounted at +26 degrees and -5 degrees with respect to nadir to generate stereoscopic image of the area along the track. Both the cameras work on the 'push broom scanning' concept using linear arrays of Charge Coupled Devices (CCDs) as sensors. The spacecraft body is steerable to compensate the earth rotation effect and to force both Fore and Aft cameras to look at the same ground strip when operated in stereo mode. It is also possible to operate the cameras in wide swath mode by suitably maneuvering the spacecraft. The spacecraft has roll tilt capability, which can be utilized for faster global coverage.

Apart from the two panchromatic cameras, the satellite carries an On-Board Solid State Recorder (OBSSR) for collecting data outside the vicinity of the ground stations. The OBSSR has a capacity of 120 GB. It can record data of 9 minutes.

Satellite Data Products

Introduction

The data products are available from different sensors of the IRS Missions. The basic parameter, which decides any product, is the Spatial Resolution, Spectral resolution, Radiometric resolution and the temporal resolution. It is also difficult to have ideal values for all these parameters in any sensor and hence it is a trade off between these parameters. The type of application decides the trade off. Hence for any process with dynamic application like Vegetation the temporal repetivity is important and in turn will have coarse spatial resolution.

The Satellite data products are defined by the need and applications. The sensor specifications are different for different missions and the primary requiremnt for the genration of any data is to convert the users area of interts in trms of the products for a spefic mission. NDC has utilitites avaible for converting the users area into specifc mission.

Specifying user area of interest

To obtain satellite data, users have to specify their area of interest. This may be

- ❖ a city
- ❖ a district
- ❖ a state
- ❖ a watershed
- ❖ any contiguous area

This can be specified in terms of

- ❖ latitude/longitudes of a polygon
- ❖ path/row number (obtained from a referencing scheme map)
- ❖ mapsheet number (as per Survey of India SOI nomenclature)

An Integrated Digital Referencing Scheme (IDRS) Software package, which can be run on any PC Windows environment, is also available. By utilising this package the user's area of interest (place, polygon, mapsheet, etc.) can be converted into path/row number of the required satellite mission. A print out of the plot showing the nominal coverage can also be obtained. IDRS package is available on the website.

Specifying Type of Product

As mentioned earlier, products can be supplied in a wide variety of media and formats. The classification of products is generally based on the following:

- ❖ Level of processing/enhancement
- ❖ Output media/scale
- ❖ Area coverage

Levels of processing

The raw data acquired from satellite has radiometric and geometric distortions. This is corrected by processing the data on computer systems. The corrected data is then put in the required format. The different levels of processing are:

- ❖ **Raw data**

  No processing is done here and the relevant scene is extracted and put on the media. Generally, available only as digital product. Useful for evaluation of data quality and for stereo data.

- ❖ **Partially processed data**

  Radiometrically corrected but geometrically uncorrected. This type of data is useful for stereo products.

- ❖ **Standard data**

  Radiometrically and Geometrically corrected data. Normally all full scene, full scene Shift Along Track (SAT), quadrant data is supplied as standard data.

- ❖ **Geocoded data**

  These products are north oriented and compatible to the SOI mapsheets. In case of photographic outputs the scale of products (1:50,000, 1:25,000) is also much better than the full scenes/quadrants.

❖ **Special data products**
Besides standard data, NRSA also offers certain special products customised to suit user requireemnts. These include mosaiced, merged and extracted data in different scales. For obtaining special products, users need to interact with NDC to specify their requirements. Some of the special products are PAN + LISS-III merged products and District/State wide mosaics. If the area of interest in large and contiguous special prices will be worked out.

Output media/scale

Satellite data products are available on photographic and digital media.

Photographic products can be supplied as films or prints in

- ❖ Black and White (B/W)
- ❖ Colour (FCC)

Generally, single band data is provided in B/W such as PAN data or one band data from multispectral sensors such as LISS-III, LISS-II/TM, etc. Similarly, photographic, colour products called as False Colour Composites (FCCS) can be provided for multispectral data.

The output of B/W FCC can be provided in the form of

- ❖ Film (negatives or positives) 240 mm × 240 mm
- ❖ Enlargements/Paperprints

    - ❑ 1X – 240 mm × 240 mm
    - ❑ 2X – 480 mm × 480 mm
    - ❑ 4X – 960 mm × 960 mm
    - ❑ 5X – 1000 mm × 1000 mm

The output scale for prints can vary from 1:1 M to 1:12,500 (for PAN point geocoded products).

For users who need the data on photographic media, the data is converted into analog form using film recorders. Since there is a change from digital domain to analog domain and to accommodate the film characteristics, an appropriate transfer function called Look-Up-Table (LUT) has to be applied during the transformation. If the LUT is achieved with the best possible enhancement of a particular scene based on its histogram, it is called scene Specific LUT. On the other hand if uniformity of color for particular feature from one scene to another is maintained using Common LUT (optimized over a wide region).

Digital data prodcuts are available on

- ♦ CD-ROMs (for IRS-1C/1D data)
- ♦ 8 mm Exabyte Tapes

The advantages of digital data are that it facilitates better spectral analysis at user end. Digital data products are more cost effective because of larger area coverage. The formats for digital data include

- LGSOWG (in BIL or BSQ)
- FAST format.

More Details on this can be obtained from the digital data products format document.

Area coverage

Different sensors have different swaths. Generally, standard products are provided as full scenes. However, to obtain better output scales or to optimize data requirements, part of the Full scenes such as Quadrants or Geocoded products or Shift Along Track products can also be provided (Table 1).

Table 1. The different types of products depending on sensor, media/scale and area covered.

Sl. No.	Type of Product	Sensors
1	Full Scenes – all bands	LISS-IV MONO, PAN, LISS-III, TM, MLA, PLA, AWiFS, WiFS, LISS-I, LISS-II, OCM, MSMR
2	Full scenes – specified bands (3 bands for FCC products)	LISS-III single band, TM single band or 4 bands, LISS-I/LISS-II single band
3	Full scenes with Shift Along Track	PAN, WiFS, LISS-III, OCM
4	Sub-scenes	PAN
5	Quadrants	LISS-III, TM, OCM
6	Geocoded Mapsheet based (15' × 15')	LISS-III, LISS-II,
7	Geocoded Floating (15' × 15')	LISS-III
8	Geocoded (7.5' × 7.5') mapsheet floating	LISS-IV MONO, MX, PAN,
9	North oriented (100 km × 100 km)	OCM
10	Full pass strip data of one array	PAN, OCM

Data Ordering

Users can place orders for data products by filling the order form specifying the sensor, type of product, period of interest, etc. alongwith 100% advance payment in the form of a DD payable to NRSA at Hyderabad, an running account will be opened at NDC and the required funds for data products deposited in the account. Upon receipt of an order, NDC will take up the same for processing and send a confirmation (in about 3 – 4 days) to the user.

Products are generally despatched within a week to ten days from the date of confirmation. All products are checked for quality before despatch. Products are generally shipped by courier/speed post.

Product Code

The important objective of any mission is the timely supply of data users. There are many varieties of products and in order to cater to the demand of the user community, NRSA had developed a unique production management system towards meeting the goals and objectives of the IRS missions. This is an automated production environment, which monitors the generation of products and ensures the timely supply of the same. In order to generate the data through this system there should not be any ambiguity in terms of products and sensor etc. So a nine-digit prduct code is defined which will cater to different combinations of products that can be generated through this production sytem. The dteials of nine-digit code and the product defenition are given in the Annexure–II.

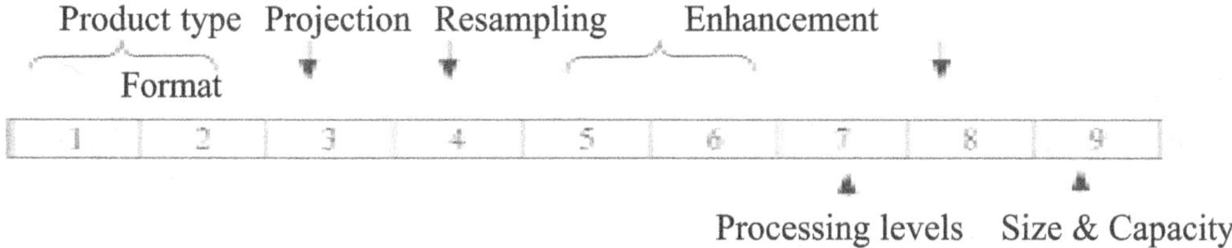

For example: STPC0026J – Standard, Polyconic, No enhancement, Systematic corrected, Digital data LOGSWOG BIL, 650 MB CDROM. This product code is same for IRS 1A or IC, and sensors LISS-I or LIS-III, PAN etc.

Thus product codes are simple and uniquely define the data from different missions that are generic in nature. However the type and sensor of the missions defines the respective product codes which is taken care in the production environment.

Payload Programming

The PAN and LISS-IV MONS MX, camera can be tilted by 26 degrees on either side to obtain more frequent revisit and to obtain stereo data. ERS-SAR data is not acquired routinely. The data requirements through OBSSR for areas pertaining to outside the visibility of Shadnagar Earth station coverage needs to be planned and informed to the respective Mission operators. The data programming for nay disaster requirement also needs programming the satellites. The above activities call for prior programming of the satellite and consequently, users who need such data should place orders in advance of their period of interest.

Foreign Satellites

Apart from the Indian Remote Sensing Satellites, NRSA acquires and distributes data from a number of foreign satellites. During the 1970's and 80's, India's remote sensing data needs were being addressed by foreign satellites like LANDSAT, NOAA, SPOT etc., where we played the passive role of just obtaining the data product. With the setting up of an Earth Station at Hyderabad in 1979, data reception started from LANDSAT satellite. Even after being equipped with a series of IRS satellites, NRSA continues to acquire or distribute data from foreign satellites to supplement the data requirements of its users. Currently, NRSA is acquiring data from NOAA-16, NOAA-17, TERRA, AQUA and ERS. Apart from acquiring NRSA also distributes data acquired by RADARSAT, IKONOS, QUICKBIRD and ENVISAT.

Archived Data

Landsat

NRSA has a huge archives of data acquired from the Landsat series of satellites during the period 1984-2001. The Landsat-4 and 5 satellites carry two sensors ob-board. A Multi Spectral Scanner (MSS) and a Thematic Mapper (TM). While MSS provided data in four spectral bands with a resolution of 80m, TM provided data in seven spectral bands with a resolution of 30m in the visible and near infra-red bands and 120m in the thermal band. As per the archival policy four coverages per year are available for day passes, two cycles in two years for night passes for the period 1984 to 1997 and data of all the passes acquired thereafter.

ERS-1/2

NRSA has been acquiring microwave Synthetic Aperture Radar (SAR) data in image from the ERS-1/2 satellites from 1992 onwards. Products can be supplied from archived data or through programming for future acquisitions. ERS data have been acquired against user requests from 1991 and Tandem data from ERS-1 and ERS-2 during 1995-96.

SPOT

NRSA has acquired Multi-spectral Linear Array (MLA) (Multispectral - 20 m resolution) and Panchromatic Linear Array (PLA) (Panchromatic 10 m resolution) data from SPOT 1 and 2 during the period 1987 to 1990.

Satellites From Which Data are Being Aquired
NOAA-KLM

Since the 1960s, National Aeronautics and Space Administration (NASA), has been developing Polar orbiting Operational Environmental Satellites (POES) for National Oceanic and Atmospheric Administration (NOAA). The latest in the series, NOAA-M, renamed as NOAA-17, the third in the KLM series of POES satellites, was launched successfully on June 24, 2002 from Vandenberg Air Force Base, California. The satellite has been placed in a sun-synchronous orbit at an altitude of 833 Km with an inclination angle of 98.6^0 with respect to the Equator. The orbital period is 101.35 minutes and equatorial crossing time is 10 A.M. The spacecraft like its predecessors carries a suite of sensors on-board.

NRSA has been providing NOAA data to the Indian user community since 1987. From 01 April 2004 onwards, NRSA is acquiring data from NOAA-17 satellite also. Currently, out of the available suite of sensors, NRSA acquires only Advanced Very High Resolution Radiometer (AVHRR) data.

NOAA-17 AVHRR is similar to NOAA-16 AVHRR except that it includes a new sixth channel in the visible range that can be used to provide the capability to distinguish between clouds and snow/ice on the ground. The six spectral bands of AVHRR are :

Channel 1	0.58 to 0.68 micrometers
Channel 2	0.725 to 1.00 micrometers
Channel 3A	1.58 to 1.64 micrometers
Channel 3B	3.55 to 3.93 micrometers
Channel 4	103 to 11.3 micrometers
Channel 5	11.5 to 12.5 micrometers

The AVHRR sensor provides data with a spatial resolution of 1.09 Km and 10 bit radiometric quantization. All six spectral channels of the sensor are registered so that they all measure energy from the same spot on the earth at the same time. Although the AVHRR has six channels, only five are transmitted to the ground at any one time. Channels 3A and 3B cannot be operated simultaneously. Channel 3A is useful for snow - cloud discrimination, while Channel 3B is useful for Sea Surface Temperature (SST) measurements. In view of the above, during day passes Channel 3A data is acquired, while during night passes Channel 3B data is acquired by default.

TERRA / AQUA

Since its creation in 1958, NASA has been studying the Earth and its changing environment by observing the atmosphere, oceans, land, ice and snow, their influence on climate and weather. In 1991, NASA launched a comprehensive program to study the Earth as an environmental system called the Earth Science Enterprise which comprised of focused, free-flying satellites, space shuttle missions, and various air-borne and ground based studies. Phase-II began in 1999 with the launch of the first EOS satellite, TERRA (formerly AM-1) and Landsat-7. The AM satellite orbits in sun synchronous polar orbit, descending southward and crosses equator in the morning. The other satellite is AQUA (PM - I). This also orbits in sun synchronous polar orbit, ascending northward in the afternoon.

NRSA has been acquiring and disseminating TERRA-MODIS data since November 2002. From January 2004 onwards, NRSA is acquiring data from AQUA satellite also. The TERRA and AQUA satellites carry number instruments out of which NRSA acquires data collected by the Moderate Resolution Imaging Spectro Radiometer (MODIS).

Moderate Resolution Imaging Spectro Radiometer (MODIS) - Moderate Resolution Imaging Spectro Radiometer (MODIS) is the key instrument in EOS Satellites. It is a passive imaging spectro radiometer. It scans across-track, a swath of 2330 km using 36 discrete spectral bands (visible, near and thermal infra red) between 0.41 and 14.2 micrometers. It provides imagery of the earth's surface and cloud cover to develop an improved understanding of global dynamics and processes occurring on the surface and in the lower atmosphere.

Some of the channels are available at different resolutions (1 and 2 at 250m, 3-7 at 500m). MODIS data comes in Hierarchical Data Format file, which enables storing of multi type data sets. The HDF file for the lower resolution data contains the higher resolution data down sampled to the lower resolution.

While the equatorial crossing time of TERRA is 10:30 A.M., the same for AQUA is 1:30 P.M. AQUA's afternoon crossing time was chosen for its usefulness for meteorological forecasting

Satellites From Which Data Are Being Distributed

RADARSAT

NRSA has signed an agreement with RADARSAT International, Canada, on 13th August 1998 for distribution of RADARSAT data products to Indian users.

Radarsat's Synthetic Aperture Radar (SAR) has the capability to penetrate darkness, clouds, rain and haze. It provides solution for acquiring data over dynamic areas like tropical, coastal and Polar Regions. RADARSAT can acquire SAR data over any part of the world. Data collected by the satellite is either directly transmitted to the local ground stations or to RADARSAT Ground receiving Station at Prince Albert, Saskatchewan or Gastineau, Quebec, Canada. The satellite is equipped with two On Board Tape Recorders to record data over areas outside the visibility of the ground stations. RADARSAT's SAR operates in C-band (5.6cm), at a frequency of 5.3 GHz with HH polarization. The horizontal polarization mode is more useful for land applications.

Radarsat provides images in seven sizes, termed "beam modes". They vary from fine beam mode (50×50 Km area in 10m resolution) to Scan SAR wide (which covers 500 $\times$ 500 Km area in 100 m resolution). Depending on the size of the area and level of details required, user can opt for any one of the 7 beam modes. RADARSAT also offers a range of incidence angles from 10 to 60 degrees, which can be used to provide more frequent revisit cycle (regular cycle is 24 days) and also offers opportunities to acquire stereo pairs.

IKONOS

NRSA is distributing IKONOS data to Indian User Community since October 2001. NRSA procures the IKONOS data from Space Imaging, U.S.A. and distributes the same to users in the government, private and academic sectors.

Ikonos provides Panchromatic data with a resolution of 1m and Multi-spectral data with a resolution of 4m with 10 bit quantization.

Space Imaging, U.S.A., acquires IKONOS data, either in real time or by using the On Board Tape Recorder. Space Imaging, U.S.A. also has many Affiliate Ground Stations all over the world, which acquire IKONOS data when the satellite passes over these stations. The Western half of India is covered by the ground station at Dubai (Space Imaging Middle East) and the Eastern half of India is under the Space Imaging, U.S.A. (SINA) International Communication Cone.

QUICKBIRD

The Antrix Corporation Ltd. / NRSA, Department of Space, Government of India, have arrived at an agreement with M/s DigitalGlobe/Hitachi Software to distribute Quickbird data to Indian Users. As in the case of IKONOS data, all Quickbird data requirements are placed with NRSA. On behalf of the User, NRSA procures the data from DigitalGlobe/ Hitachi Software and disseminated.

Users should place the indents with NRSA in the prescribed high-resolution data request form with the required undertakings as per Government guidelines.

The QuickBird satellite collects both multi-spectral and panchromatic imagery concurrently. Pan-sharpened composite products in natural or infrared colours with a resolution of 70 cm are derived from the above. Panchromatic imagery is collected in 11-bit format (2048 gray levels) and delivered in 16-bit format for superior image interpretation (shadow detail, etc.), or 8-bit format (256 gray levels). The panchromatic sensor collects information at the visible and near-infrared wavelengths with a bandwidth of 450-900 nm. Four-band multi-spectral imagery consists of blue, green, red and near-infrared bands, delivered in 16-bit and 8-bit formats. In addition, Pan and Pan-sharpened Imagery are available at 60 or 70-centimetre resolution.

Both strip mode and area mode imaging are supported. Strips up to 10 scenes in length (165 km) can be collected in a single pass. Areas up to 2x2 scenes can be collected in a single pass. QuickBird's on-board tape recorder gives world-wide coverage.

While DigitalGlobe/Hitachi Software offer Imagery products at three levels namely, Basic Imagery, Standard Imagery and Ortho rectified Imagery, it is proposed to distribute only standard corrected, geocoded data to Indian Users. It is also possible to obtain stereo data from the satellite.

ENVISAT

NRSA has been distributing RADARSAT data to Indian users to meet their microwave data requirements. NRSA now has commenced distribution of European Space Agency's ENVISAT data. This satellite was launched on 1st March 2002.The satellite carries a number of instruments, which aid in land, oceanographic and atmospheric studies. NRSA will distribute only Advanced Synthetic Aperture RADAR (ASAR) data from this satellite. This sensor provides all weather - day and night RADAR images.

Future IRS Satellites

Encouraged by the successful operations of the various IRS missions, a number of missions are planned for realization during the next few years. These missions are designed to carry suitable sensors for applications in cartography, oceanography and atmospheric studies and microwave remote sensing.

CARTOSAT-2

This satellite will be placed in a polar, Sun synchronous orbit at an altitude of 630 km and alocal mean time - 9.30 A.M. This satellite will carry a single PAN camera with high agility.

The PAN camera collects data in 0.45 - 0.85 mm band with a spatial resolution of 1m and 9.6 km swath. The camera will provide data with 10-bit radiometric resolution.

The camera can be operated in the following modes

- **Spot mode of Imaging :** Strips on the either side of the track in the North-South direction.

- **Paint brush mode of Imaging :** This mode is used to increase the total swath. Both Roll tilt and pitch tilt is employed.

- **Multi-view mode of Imaging :** Strips on the either side of the track in the North-South direction are considered and one strip selected is imaged three times with different look angles.

- **Imaging in South-North Direction:**
    - to avoid the imaging of shadows regions.
    - for Certain Sun declinations, Image trace direction has to be opposite to the nominal forward direction.

Radar Imaging Satellite (RISAT)

This will be the first satellite to be launched by India to operate in the microwave region. The satellite will carry a multi-mode, agile Synthetic Aperture RADAR (SAR) payload operating in ScanSAR strip and Spot modes to provide images with coarse, fine and high spatial resolutions respectively.

OCEANSAT-2

This satellite mission is conceived to provide continuity of services to the Oceansat-1 data users. This satellite will have enhanced capabilities. It will carry an Ocean Color Monitor (OCM) and Wind Scatterometer. Inclusion of a thermal infrared Radiometer is also under consideration.

Conclusion

NDC has a range of Satellite data products with a wide choice of resolutions, processing levels, product media, output scales, area coverage repetivity, season and spectral bands. The IRS missions and the products provide means to target specific sites of interest in MONO as well as Multi spectral bands. The option of stereo viewing enables the users to get the Stereo data and in turn Digital Elevation Models .The tiltabilty and flexible viewing options have the potential for short revisit cycles and therefore rapid update of information

Annexure - I

Table 2. Information of IRS Bands, Resolution, Repetivity, Coverage.

SAT	SEN	Resolution (M)	No. of Bands, Region	Revisit (DAYS)	Coverage (KM x KM)
IRS1A	LISS-I	72.5	Four		148 x 174
	LISS-II	36.25	0.45 - 0.52	22	74 x 87
			0.52 - 0.59		
			0.62 - 0.86		
IRS1B	LISS-I	72.5	Four		148 x 174
	LISS-II	36.25	0.45 - 0.52	22	74 x 87
			0.52 - 0.59		
			0.62 - 0.86		
IRSP2	LISS-II	36.25	Four	22	74 x 87
			0.45 - 0.52		
			0.52 - 0.59		
			0.62 - 0.86		
IRSP3	WIFS (V_NIR)	188	Three		
		188 x 246	0.62 - 0.68	24 *(3**)	810 x 810
			0.77 - 0.86		
	WIFS (SWIR)		1.55 - 1.69		
IRS1C			Three	24*(5**)	70 x 70
	PAN	6.25	0.50 - 0.75		141 x 141
	LISS-III	23.5	0.52 - 0.59		810 x 810
			0.62 - 0.68		
			0.77 - 0.86		
			1.55 - 1.69		
	WIFS	188	0.62 - 0.68		
			0.77 - 0.86		

Table 2. *Contd...*

SAT	SEN	Resolution (M)	No. of Bands, Region	Revisit (DAYS)	Coverage (KM × KM)
IRS1D	PAN	5.2-5.8	Three	25* (3**)	63 × 70.8
		21.2-23.5	Same as BC		70 × 69.18
	LISS-III	184-188			127 × 149.3
					141 × 142.1
	WiFS				720 × 778
					912 × 760
IRS P4	OCM	360 × 250	Eight	2	1420 × 1420
			0.402 - 0.422		
			0.433 - 0.453		
			0.480 - 0.500		
			0.545 - 0.565		
			0.660 - 0.680		
			0.745 - 0.785		
			0.845 - 0.885		
IRS P6	LISS-IV (Mono)	5.8	Three	24* (5**)	70 × 70
	and	5.8	0.52 - 0.59		23 × 23
	(MX)		0.62 - 0.68		
			0.76 - 0.86		
		23.5			142 × 141
	LISS_III		0.52 - 0.59		
			0.62 - 0.68		
			0.76 - 0.86		
			1.55 - 1.70		
		56 (NADIR)	0.52 - 0.59		
	AwiFS	70	0.62 - 0.68		736 × 737
		(read pixel)	0.76 - 0.86		
			1.55 - 1.70		
IRS P5	PAN	2.5	0.50 - 0.80	126* (5***)	30 × 30

Annexure-II

The details of the different product types and related information for defining the products.

Product Type (First two characters)

ST   Standard product

QU   Quadrant product

G3   Geocoded product (15' × 15' of SOI mapsheet)

G4   Geocoded product (7.5' × 7.5' of SOI mapsheet)

SR   Stereo Pair product

TR   Shift along the track product

J3   Geocoded product (15' × 15' without SOI reference)

J4   Geocoded product (7.5' × 7.5' without SOI reference)

J5   Geocoded product (5. × 5' without SOI reference)

Projection Applied (Third character)

  O No projection applied
  P Polyconic projection
  S SOM projection
  L Lambert projection

Resampling used (Fourth character)

  O No sampling done
  C Cubic convolution
  N Nearest neighbour

Enhancement (Fifth & Sixth character)

  00 Enhancement (mostly applicable for digital)
  01 Histogram Look Up Table (Scene based enhancement)

Levels of processing (seventh character)

  0. Raw data
  1. Partially corrected data
  2. Standard product

Format (Eighth character)

  1 Block & White positive
  2 Black & White pnnt
  3 False colour composite film negative
  4 False colour comoosite film positive
  5 False colour composite paper print
  6 Digital data LGSOWG, Band interleaved (BIL)
  7 Digital data on LGSOWG, Band Sequential (BSQ)

Size and capacity (Ninth character)

  2 240 mm (only for photographic products)
  3 500 mm (only for photographic products)
  4 960mm (only for photographic products)
  5 1000 mm (only photographic products)
  6 1600 BPI (only for digital products)
  7 6250 BPI (only for digital products)
  H 525 MB Unix based cartridge
  I 5 GB 8mm Exabyte DAT
  J 650 MB CD-ROM

Securing Enough to Eat : Role of Space Technology

S Bandyopadhyay, SK Srivastava, VS Hegde and V Jayaraman

EOS Programme Office, Indian Space Research Organisation Headquarters, Bangalore- 560 094

ABSTRACT

The ever-growing population necessitates the increasing demand of food production in the next few decades. Amidst the limited availability of land and water system with enormous human pressure, exploitation of agro-ecosystem for enhancing the food production necessitates - increasing the productivity on sustainable basis. Therefore, optimal strategies need to be framed with proper blending of core, enabling and facilitator technologies in order to achieve the necessary crop productivity of our agricultural ecosystem involving land and water. In this context, space remote sensing technology has been proved to be the potential tool in providing the timely and accurate information for decision making while inventorying the natural ecosystem. With constellation of Indian Remote Sensing Satellites in orbits and unique capabilities of IRS P6 & P5, it is being possible to create efficient natural resource database to address the issues related to food production. While the infrastructure created in space has got enough potential in terms of capturing the status, vulnerability and dynamism of agricultural eco-system, the capacity is yet to harness fully. The gap between technological improvement, with regard to space remote sensing and its operational utilization down the line in agricultural sector has to be brought down. Towards this, the present paper highlights the role of space technological inputs in expanding cultivable lands, enhancing irrigation and water use, enabling intensification/ diversification, participatory watershed management, integrated pest management, risk reduction through drought management, enabling micro-credits, crop insurance, agro-meteorological services, informatics for agriculture - agricultural statistics. The paper also outlines the strategy how space applications in agriculture sector would drive India's efforts in securing enough to eat.

Introduction

In a predominantly agricultural country like ours, sustaining and strengthening agricultural progress is of prime importance. The food security incorporates not only the traditional idea of ensuring adequate food *availability*, but also the need to create the social and economic conditions which empower households and individuals to gain *access* to food, either by producing food themselves or earning income to buy food. Finally, food security assumes the effective and efficient *utilization* of food. Efforts to promote long-term food security, therefore, will include a wide array of measures aimed broadly at increasing production, eradicating poverty, improving health and nutrition, and empowering women as both food producers and caregivers (Swaminathan, 2001).

Estimates made in the eve of the 21st century indicate that, by the year 2020, India may have 1.4 billion people - a little less than 150% of 1.0 billion population count reached during 2001. To continue to keep the present per capita food grain availability of around 500 gm/person/day (Directorate of economics and statistics, 2004) the total food requirement would be 255.0 million tones by 2020 (Figure - 1). This calls for increasing the overall average productivity from the present 1.7 tonnes/ hectare to at least 2.1 tonnes/ hectare from average net sown area of 122 million hectare out of estimated 142 million hectare. Further, productivity improvement in dryland agriculture is of major challenge. The productivity gap between the rainfed and irrigated area is of much concern as 41% of net sown area is irrigated (Yadav et al., 2000). The per hectare food grain production from irrigated area is 2.5 tonnes (40% of the realizable potential) as compared to just 1.0 tonnes from rainfed area.

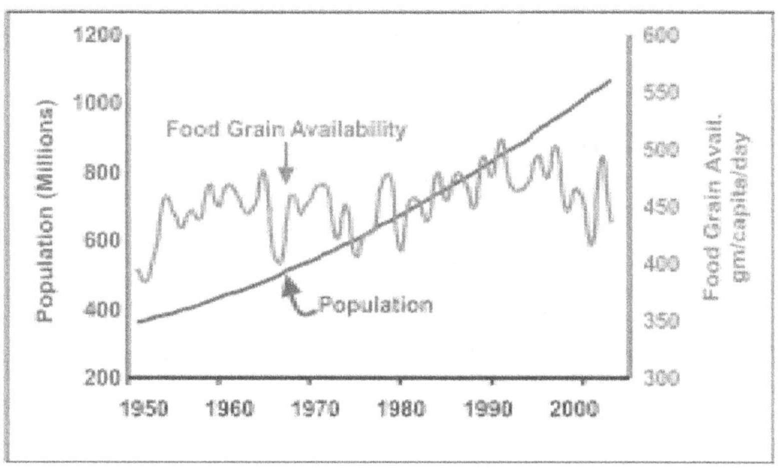

Fig. 1 Increase in population and food grain availability over the last five decades

Indian agriculture has stride through the hard times in the past under the various colonial regimes. In the post independence era during 1949-50 the food grain production was as low as 55.0 Million Tones to feed around 350 million people with productivity rate of mere 0.55 tones/hectare. After five decades, the population has reached 1000 million with average annual growth rate of 13 million people and at the same time food production touched 212.0 million tones with average productivity of 1.6 t/ha. Thanks to our technological breakthrough in the field of crop sciences and by developing protein rich animal products.

The impressive increase in food grain production during 1950 to 2004 is attributed to the progress made in various sectors related to farming viz., introduction of high yielding varieties replacing the low productive traditional varieties; increase in consumption of fertilizer from 0.6 Mt to 17.0 Mt; increase in net irrigated area from 21.0 Mha to 55.0 Mha (ie., 17% to 38%); increase in cropping intensity from 93% to 133% in 1999; boost in use of pesticides from very nominal amount to 0.85 Mt; increase in net sown area from 119 Mha to 142 Mha (Economic Survey of India, 1999-2000).

Today we are at the crossroad. The green revolution, the country had experience in the late sixties, left profound impact on Indian agriculture. Although the productivity has steadily progressed and reached three times in last 50 years, but it remains stagnant during last one decade creating a plateau. This is become cause of concerns in the front of food security. The rainfall dependency, marginalized and poverty stricken farmers, inefficient input management and over all knowledge poor farming society is struggling with hand to mouth sustenance by low return from the marketable surplus. Therefore, the urgent need is to break the productivity doldrums with technological interference and adequate strategies to feed the ever-increasing population in the coming days.

The optimal amalgamation of core, enabler and facilitator technologies is necessary to address the issues of agro-ecosystem in 21st century (Figure - 2). Space technology, in recent years as facilitator has proved to be the imperative tool for mapping, monitoring and management of agro-ecosystem viz., land use, crop productivity, soil & water resources, environment, urban sprawl and flood & drought. In India, constellation of IRS satellites, very recently - IRS-P6 and IRS P5 have given a boot to satellite base resource inventory. Remote sensing primarily records electromagnetic energy that emanates from the earth surface. Specific or combinations of various spectral bands in the range of electromagnetic spectrum are being utilized for collecting information about the earth resources. While the temporal resolution of the remote sensing satellites enable monitoring the dynamic features like agricultural system, the high spatial resolution provides accurate information about crop extent and diversity.

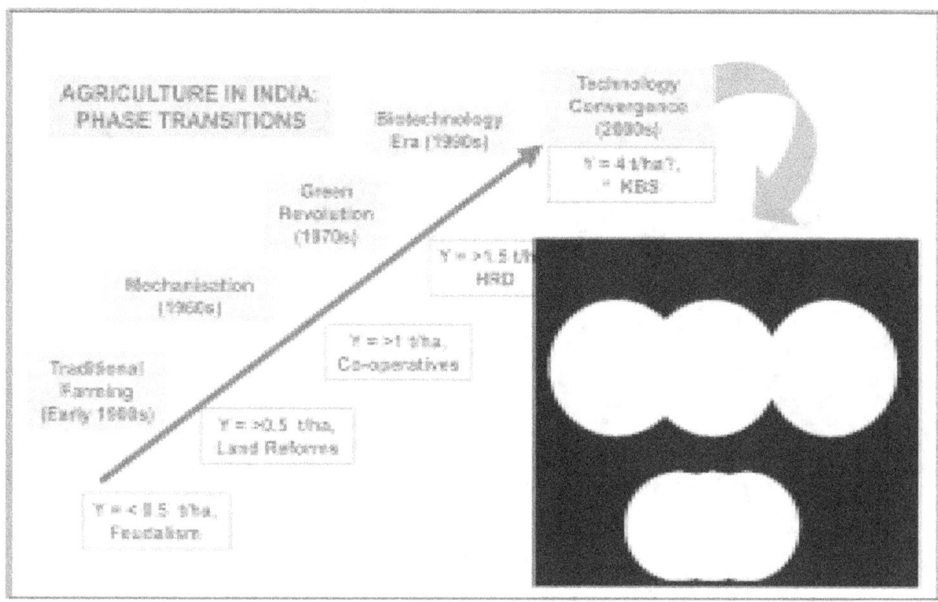

Fig. 2 Technology Convergence in Agriculture

In India, amidst enormous climatic diversity and colossal anthropogenic pressure the state of natural resources, particularly, soil and vegetation is at stake. Combining the crucial information provided by space remote sensing with meteorological information and crop genetical characteristics, the Indian remote-sensing scientists achieved the pathway for integrated sustainable management strategy during the last decade. The information from space remote sensing in combination of allied information are being utilized for implementing the measures for conservation of soil and water at watershed level and improving the agricultural productivity. In the context of food security, constant monitoring from space acts as major catalysts in ensuring the sustainability of the development strategy and the maintenance of the agro-ecosystem, thus creating the new paradigm (Figure 3). The paper describes the advances in space technological applications in ensuring the environmental integrity while mechanized exploitation of land and water resources to make the food accessible to all.

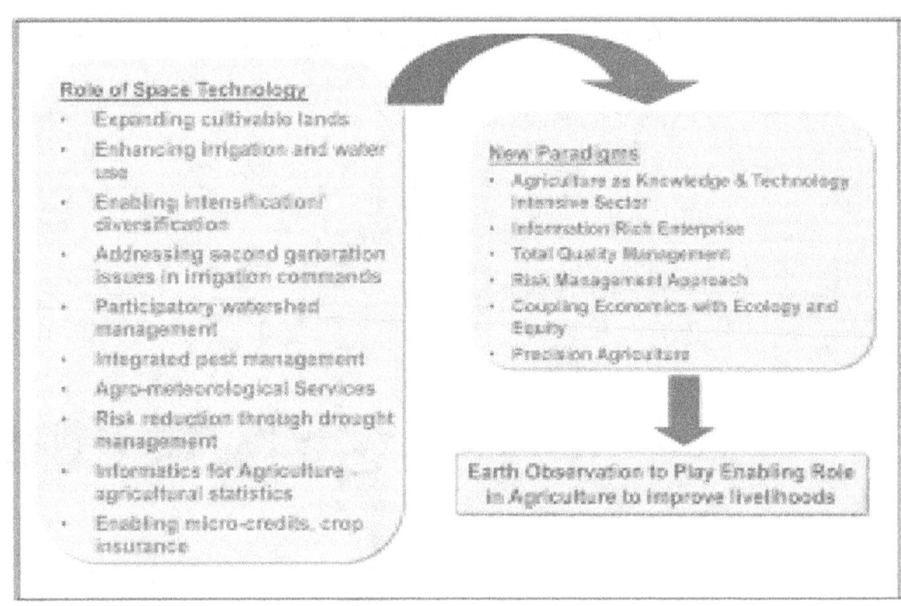

Fig. 3 Earth observation inputs towards food security

Issues and Perspective of Food Production in India

Sustained growth in agricultural productivity is critical to improvements in food security. The growth in agricultural productivity translates into increased food availability and lower food prices for consumers. The growth in agricultural productivity means higher incomes for many small and marginal farmers who earn their livelihoods through agricultural production. Most changes in agricultural productivity over time (and differences across farms or countries) can be attributed to differences in the quantity of resources used in agricultural production, such as land, labor, and fertilizer. Distinguishing the relative impacts of resource quantity and quality is important in determining appropriate policy measures to improve agricultural productivity and food security.

In India, historically, Agriculture has seen three different phases of technological era viz., mechanization, green revolution and biotechnology. While the productivity levels in developed countries (mostly European and US) have gone up from 1.5 t/ha to around 4t/ha, world average is just touching 2t/ha (FAO, 2001 & 2002). Genetic revolution, initiated in the 1960s, was primarily based on high technology package involving large-scale use of chemical fertilizers and pesticides, better yielding seeds and extensive irrigation. While developed and lately Asian countries largely realized the benefits of green revolution, African countries are still far behind (Figure 4). The dramatic increase in India's food gain production is clearly demonstrates the impact of green revolution.

The impact of technological changes on agriculture system has been of three types : expansionist, exploitative and destructive. Expansionist nature was witnessed between 1950 and 1970 (just before green revolution period) and arable land was increased from about 95 Mha to 130 Mha. Agriculture activities were expanded horizontally by bringing more land under plough. Green revolution has turned agriculture more intensive by exploiting moisture and nutrients from the soils by high yielding crop varieties, while arable land marginally increased (130 Mha to 140 Mha), productivity went up three times (more than 0.5 t/ha to 1.5 t/ha). Destructive phase started with (i) irrigated command areas turning saline/water logged and (ii) rainfed areas witnessing massive soil erosion. This is, in fact, destruction of agro-ecosystems and this trend has to be reversed by turning agricultural practices regenerative. This is where the concept of sustainable agriculture holds the key.

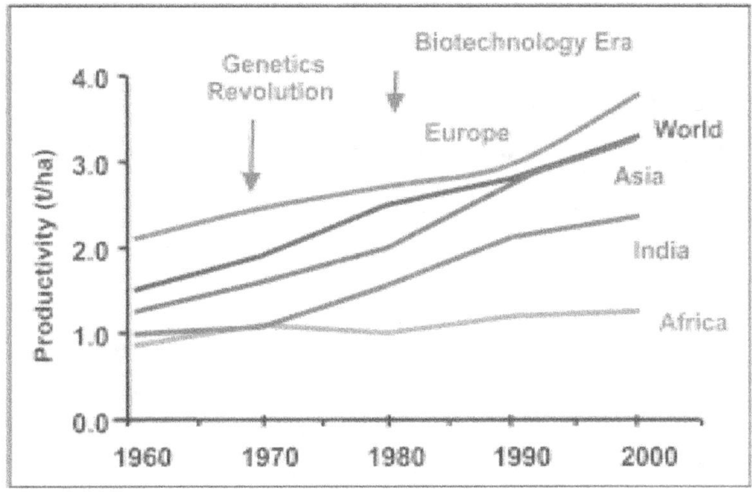

Fig. 4 Food productivity trends of World, Asia, Africa and Europe

Introduction of high yielding crop genotype coupled with inefficient and indiscriminate use of water, inorganic fertilizers and insecticides/pesticides has resulted in the disruption of agro-ecosystems. It is reflected from the fact that the severity of pest diseases problem in tropical countries is 5-10 times more as compared to temperate climate. As the result of this vulnerability, the estimated crop loss due to pest alone is about 20% of potential production in developing countries. The problems of dry land agriculture are characterized more in terms of soil erosion and desertification (Figure - 4). These areas offer tremendous potential towards boosting food production, if the strategy for food production is coupled with in situ conservation soil and water. Overexploitations of dry lands, recurrent droughts and large-scale deforestation have resulted in severe degradation of arable land and increasing desertification all over the world. Soil erosion rates subsequent to unscientific agricultural practices have increased to such an extent that the material delivery from rivers to the oceans has been increasing which result in frequent flooding.

Water use efficiency has always been an important issue associated with food production in India. Four monsoon months contribute 300 Million hectare metre (Mham), while remaining 8 months contributes just 100 Mham. Around 10 Mham surface water is contributed by snowfall, while 20 Mham is contributed from the waterways having the origin outside India. Today, India uses only a tenth of the rainfall it receives annually and even 25 years from now, will be using only a quarter. But it must learn to store the water and use even the fraction it uses without polluting it; otherwise there will be serious water shortages. India's groundwater resources are almost 10 times its annual rainfall. But with over 200,000 tube wells added every year, the water table is declining in many areas, leaving the dug wells of the poor high and dry.

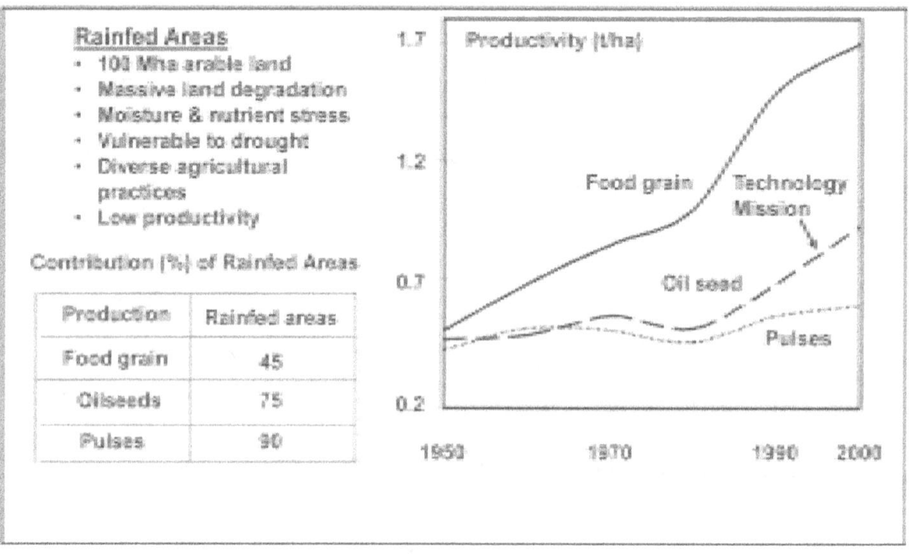

Production	Rainfed areas
Food grain	45
Oilseeds	75
Pulses	90

Fig. 5 Scope of enhancing food productivity

In general, the greatest improvements in agricultural productivity will be realized from addressing the above constraints and management factors to make the differences in the years to come. It is not surprising that the quality of soils and climate should play a key role in defining these differences. Yet only recently, with improvements in spatial data and methods, has it become possible to characterize these differences with increased precision on a regional scale. Continued research will further refine our understanding of the links between resource quality, agricultural productivity, and food security - and of the tools that policymakers can use to improve them.

Space Remote Sensing – Physical Aspects

Characteristic signatures of crops, extracted from its unique reflectance spectra that are obtained from space-borne spectroscopy, form the basis of remote sensing of crops. Taking into account the concept of reflectance and emission spectroscopy, the crops are classified based on the consistency of its signature patterns in object and image domains. With the developments related to high resolution imaging sensors coupled with robust classifiers, the satellite information technology, through application of remote sensing, has sown tremendous progress during the last two decades. By virtue of having high spectral, spatial, temporal and radiometric resolutions, remote sensing offers unique potentials to address both the structural and functional attributes of agricultural ecosystems more appropriately, including the fragility of dry land agricultural systems (Rao, 1995).

The usefulness of the visible and infrared portions of the spectrum can be seen by comparing the typical spectral responses of a crop canopy and a bare soil. The measurements of reflectance in the red and NIR can be used to determine differences in crop canopy densities. The response of vegetation in the red and NIR have been used to form "Vegetation Indices," which typically involve some ratio of near infrared to red reflectance which find various applications in the crop growth assessment.

The research on active and passive microwave remote sensing over the last two decades has confirmed that microwaves are sensitive to both soils and crop characteristics. Microwave backscatter (using synthetic aperture radar) can detect differences in crop type, crop growth stage and crop indicators like crop height, biomass and leaf area index. The multi-beam modes associated with recent microwave satellites (eg., RADARSAT) also provide significant flexibility related to the timing, spatial resolution and incidence angle of the acquired imagery. Analysis of airborne polarimetric SAR imagery suggests that radar will be able to provide some within field information. Linear cross-polarizations and circular polarizations appear to be particularly sensitive to crop and soil conditions. However, for larger biomass crops like corn, saturation of the SAR signal will limit the use of radar imagery for monitoring the condition of corn crops to early in the growing season   (McNairn et al., 2000).

Another area of the spectrum that is useful for assessing crop conditions is the thermal portion of the spectrum. Measures of radiance in this area can be used to derive the surface temperature of the crop. As the water transpires from the plant, its leaves are cooled. In absence of enough water, its surface temperature will increase which is an indicator of it water status – crop water stress. (Jackson et al., 1986)

Yet another area of sensing is hyperspectral imaging techniques developed with the aim of primary contribution in exploring and developing new applications through the selection of optimal spectral band parameters (bands position and widths). However, in most applications, practical operational considerations (sensor cost, data volume, data processing costs, etc.) will favor the use of multi-spectral systems. Hyperspectral sensors will find a use in general purpose instruments such as spaceborne systems that provide data to a broad range for agricultural ecosystem. Hyperspectral remote sensing has the potential to identify and measure many scene features important to agriculture including, specific carotenoids, particular weed species, crop water status, crop varieties, soil organic matter, soil texture, and soil carbonates, as well as delineate management zones (Thenkabail et al., 2000).

Building Information Base - Role of Space Technology
In India, acreage and production forecasting of certain crops together with crop yield modeling and crop stress detection are operationally being carried out using remote sensing satellite data. Crop acreage estimation has been quite successful through both visual and digital analysis in many areas. However, a major constraint is cloudy sky in the kharif season when normal optical remote sensing unable to acquire good data. With the emergence of microwave remote sensing data, this difficulty can be overcome as such instrument can see through cloud.

Apart from generating agricultural statistics, remote sensing is a powerful monitoring tool. Various factors such as intensive cultivation, monocropping, mechanization and indiscriminate use of fertilizer and pesticides frequently resulted unfavourable conditions for crop growth (pest attack, nutrient toxicity etc.,) leading to the huge crop loss. Minimizing such losses through crop monitoring is one of the ways to increasing crop production, and remote sensing as a tool found to be very useful in frequent monitoring and assessing the large areas under cultivation. Besides the remote sensing satellite, the constellations of INSAT - series satellites provide the vital information on periodical weather fluctuation affecting the agro meteorological condition of the crops.

Informatics for agriculture - Agricultural Statistics
A good knowledge of cultivated areas and agricultural production is indispensable to major policy decisions concerning development plans, regional planning, trade-balance management and food-distribution management. Satellite imagery is being used for geographical stratification (sub-division in homogenous land units), area sampling frame, regression estimate and area sampling survey. The Crop Acreage and Production Estimation (CAPE) project and Forecasting Agricultural Output using Space Agrometeorology and Land-based Observations (FASAL) is being launched for the multiple productions forecasting of the major crops with improved accuracy, timeliness and scope.

In addition to the low productive dry lands, there are some more critical issues viz., land degradation, drought & floods, crop pests & diseases, nutrient loss, post harvest losses and inappropriate agricultural practices, constraining the cause of food security in the country. Managing these constraints covers a wide gamut of activities ranging from the agricultural research and appropriate package of scientific practices, basic inputs of seed, fertilizers, micro-nutrients, plant protection, credit- both short-term and medium/long term, soil moisture, transmission of research results to the field to post harvesting technology and marketing. It is here that the relevance and utility of informatics in agriculture assume paramount importance. Towards these, a national infrastructure, called National (Natural) Resources Information System (NRIS), has been established for the availability of organised spatial and non-spatial data and multi-level information networking to contribute to local, national and global needs of sustainable development. Similarly, single window approach based pilot project to supply information to all the users and decision makers at the grass root level farming community is being implemented in the form of Agro-climatic Planning and Information Bank (APIB). One of the most crucial information that is directly linked up with pricing, export-import, procurement etc., is the pre-harvest production estimates of the major crops in the country. All these efforts are aimed at building the informatics infrastructure in the agriculture sector.

The application of crop simulation models in assessing the potential production in various agro-ecosystem has been well established. Further, coupling of spectral inputs (e.g., expressing leaf area index in terms of NDVI) into these models helped in addressing the several uncertainties and developed a better understanding of biospheric interactions (Figure 6).

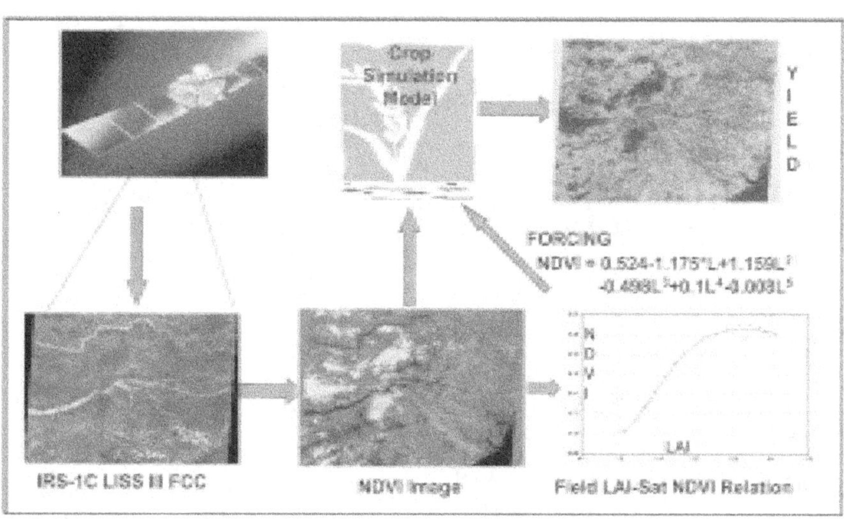

Fig. 6 Spectral interface into crop simulation model

Risk Reduction: Agriculture Crop Insurance
Implementing agricultural risk reduction strategies, including crop insurance, calls for understanding the insights on the crop risk, as a function of agro-ecological factors, cropping system, social profiling etc. The knowledge from such understanding holds significance to

develop the operational techniques towards implementing the various crop insurance schemes viz., National Agricultural Insurance Scheme (NIAS).

The developments in Earth Observations (EO) and modeling techniques provide insurers valuable dynamic information support like in season crop acreage, stress, damage and production assessment. Besides this, it also provides the inherent agro-ecological and farming system perspectives. These knowledge inputs may enable better risk reduction strategies (Figure - 7). In case of crop insurance, for example, these inputs may help in improving efficiency in operations of crop insurance scheme through timeliness and objectivity of the information, which also may lead to the reduced administrative and operative expenses and enhanced coverage of crops as well as areas.

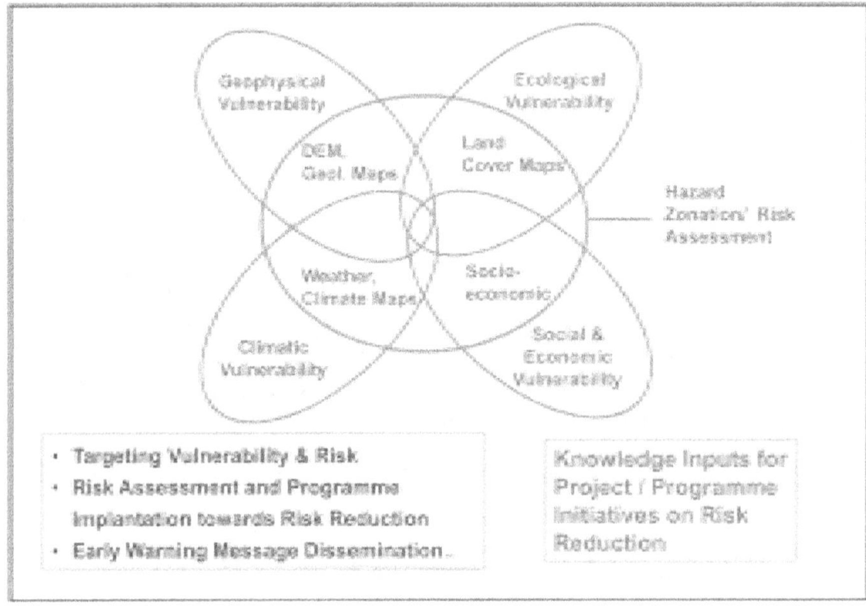

Fig. 7 Risk assetessment and risk reduction strategies

Enabling intensification & diversification

Since, land and water are shrinking resources for agriculture, there is no option except to produce more food and other agricultural commodities from less per capita arable land and irrigation water. In other words, the need for more food has to be met through higher yields per units of land, water, energy and time. It would therefore be useful to examine how science can be mobilised for raising further the ceiling to biological productivity without associated ecological harm. It will be appropriate to refer to the emerging scientific progress on the farms to emphasise that the productivity advance is sustainable overtime. Under the initiative of 'Cropping System Studies using Remote Sensing and Geographical Information System' - it has been demonstrated successfully the crop intensification and diversification in different agro-climatic regions taking into consideration of sustainability issue.

Participatory Watershed Management

Participatory watershed development programmes in many states use remote sensing data in an operational manner for soil and water conservation. For e.g., the World Bank aided Sujala Watershed Development Programme in Karnataka undertaken by the Watershed Development Department, Government of Karnataka use remote sensing and GIS for monitoring and evaluation of 77 sub watersheds in Kolar, Tumkur, Chitradurga Dharwad and Haveri. Similar programmes are ongoing in many other States. The Integrated Mission for Sustainable Development (IMSD) programme carried out covering 175 districts of the country representing diverse terrains, agroclimatic zones and social practices. Site-specific solutions have been evolved according to the land capability assessment made through integrated resource surveys using satellite based and conventional data. The IMSD project was basically aimed at evolving the scientific solutions in terms of the action plans for development of land and water resources. It is important to note that IMSD experiment was conducted in the terrain of low productivity, mainly dry lands/rain fed areas predominantly wastelands, where the scope of enhancing food production is the main issue. Under the National Watershed Development Programme for Rainfed Agriculture, (NWDPRA) of the Ministry of Agriculture, the monitoring and evaluation of watershed treated under the programme is done using remote sensing and GIS. Such monitoring and impact assessment has been carried out for many watersheds distributed in the states of Maharastra, Madhya Pradesh, Orissa, Rajasthan, Uttar Pradesh and Tamilnadu.

Irrigation Water Management

Management of water supplies for irrigation in command areas requires information on total demand and its distribution. The current satellite remote sensing capabilities for irrigation water management include end-of-season evaluation of canal command areas at the disaggregated level and diagnostic analysis of problem distributaries to enable follow-up corrective management.

Satellite remote sensing had been applied for base line inventory of irrigated area, cropping pattern, crop condition and productivity in irrigation systems, monitoring irrigation status through the season, optimum design of crop cutting experiments etc.

Simultaneous monitoring of crop performance through the irrigation season is a critical element in effective water management in the irrigation projects. Monitoring of crop condition is essential to ensure that the irrigation requirements are adequately met by canal releases. The 5 days repetitive coverage of WiFS sensor provides the necessary surveillance capability across the command area through the season. The inventory capability is significantly enhanced with the higher spatial resolution offered by LISS III sensor and the additional SWIR band would improve crop condition assessment. WiFS sensor helps in concurrent monitoring during the irrigation season, generating near real-time information on the sowing progress and crop condition for appropriate irrigation water delivery. The stereo coverage from PAN sensor providing contour maps of 10 m interval or better, helping in preliminary analysis of canal alignment and land/infrastructure development. In addition, PAN data of 5.8 m resolution is expected to be useful in delineating details of canal network and field level irrigated crop inventory

The improving spatial, spectral and temporal resolution of Indian Remote Sensing satellites has lead to significant contributions to irrigation management, ranging from simple inventorying applications to concurrent monitoring and management. The future IRS satellites will further enhance the satellite remote sensing potential in sustainable irrigation management.

Soil Salinisation and Water Logging
The development of soil salinity in irrigated agriculture has become more and more damaging to the productive lands. Breeding of plants to tolerate higher salt levels is helpful, but not the answer. Salination affects the lands that have been irrigated to realize their productive potential for crops, mainly in the regions with high evaporative demands and low water table depth. The quantification of the problem requires thorough inventory of land resources and recommendations of optimal ameliorative measures. Many irrigation projects in India and elsewhere suffer from adverse impacts of excessive irrigation and poor drainage. Satellite remote sensing has helped in mapping areas under salinization and waterlogging. In many irrigation projects in India, the areal extent and severity of waterlogging and salinisation have been mapped. The severity of such soil limitation can be either directly detected through remote sensing or seen by the impact on crop productivity.

Drought Assessment and Monitoring
The periodic droughts are very common in India due to failure or irregular onset of monsoon as the Indian agriculture is mostly dependent on rainfall with an estimate of around 94 million ha rainfed area out of 142 million ha arable land. Therefore, the drought combating strategies like drought prone area programmes have been taken up and showing encouraging results. In this context India has developed a national/regional system for assessment and monitoring of agricultural drought using satellite data base vegetation index to provide timely information on drought severity. Towards this, National Agricultural Drought Assessment and Monitoring System (N-ADAMS) have been launched in 1987 to characterize the severity of drought in the country. Since then, N-ADAMS Bulletins on monthly and seasonal crop conditions depicting agricultural drought condition are issued at States/District levels, based on vegetation indices and ground based information for 14 states and sub district levels for 6 States during the kharif season.

Conclusions

The concept of food security is a time dependent variable. After achieving the break through in food production by developing high yielding varieties and better inputs managements including irrigation, fertilizers and pesticides – the green revolution, still the India's growth in productivity sector remains low as compared to the world average. In the coming decades the major challenges lies in the realising vertical growth in food production as horizontal growth has already reach to its climax. To convert the *green revolution* to *ever-green revolution* – a continuous improvement in crop and animal productivity need to be ensured in tandem with ecological integrity. It will lead to growth in productivity in perpetuity.

Keeping in mind the tenth plan period thrust areas among others - in achieving 4% growth rate in agriculture and 8% in horticulture, animal products and fisheries, widening the food and feed security baskets and unleashing the untapped potential of major farming systems through integrated packages of technologies, services, including inputs like seed and public policies – the major focused areas are waste land & watershed development, soil health management, creating organic farming zones, integrated pest management, water management, deep sea fishing.

Towards this, space remote sensing techniques provides enough opportunity in developing water resource information system. A number of case studies on command area development, ground water inventory, canal alignment, irrigation performance evaluation, potential fishing zone forecasting etc have proved beyond doubt that integration of remote sensing and conventional approach significantly decrease the cost and time involved as well as, improve the steadfastness. Synergic use of remote sensing with the recent advance technologies such as GIS and GPS fine-tunes the spatial databases in acquiring the required locational information on the ground and analysis of classified maps. The contemporary data from the IRS satellites in orbits provides unique opportunity towards comprehensive monitoring of land and water resources dynamics. The forthcoming satellites with high radiometric, temporal and spatial resolution would help in optimised the management plans in minute details.

As the scope for expansion in area is limited - we have no option except to produce more crops per drop of water and per plot of land. Other than productivity - revolutions are also must in the sectors of quality, income and livelihood, management and marketing to put the Indian agriculture in forefront and to make it economically rewarding. This calls for suitable technological interventions and providing livelihood security to the vast rural mass engaged in agricultural business. Strategic research involving frontier technologies, such as bio, information, space, nuclear and renewable energy technologies needs considerable intensification. The information revolution thus created will help in modifying the Indian agriculture as a knowledge intensive farming system.

References :

Directorate of economics and statistics. 2004. Agricultural statistics at a glance. Department of Agriculture and Cooperation, Ministry of Agriculture, Government of India, New Delhi.

Economic Survey of India. 1999-2000. Ministry of Finance, Government of India.

Food and Agriculture Organization of the United Nations (FAO). "FAOSTAT." 2001 & 2002. http://apps.fao.org/default.htm.

Jackson R.D., Pinter Jr. P.J., Reginato R.J., and Idso S.B. 1986. Detection and Evaluation of plant stresses for crop management decisions. *IEEE Trans. on GeoSci. and Remote Sensing* GE-24(1):99-106.

McNairn H., Brown R. J. , McGovern M., Huffman T., Ellis J. 2000. Integration of Multi-Polarized SAR Data and High Spatial Optical Imagery For Precision Farming; Proceedings of the 22nd Canadian Symposium on Remote Sensing, Victoria, B.C., August 21-25, 2000.

Rao, U. R. 1995. Space technology for sustainable development, 25. Tata McGraw-Hill Publishing Company, New Delhi.

Swaminathan M. S. 1999. A Century of Hope: Harmony with nature and freedom from hunger, In East West Book (Madras) Pvt Ltd.,Chennai.

Swaminathan M.S. 2001. Food Security and Sustainable Development. *Current Science.* Vol. 81 (8): 948-954.

Thenkabail P., R. Smith E. Pauw. 2000. Hyperspectral Vegetation Indices and their Relationships with Agricultural Crop Characteristics. *Remote Sensing of Environment* 71:158-182.

Yadav, R.L., Singh, S.R., Prasad, K., Dwivedi, B.S., Batta, R.K., Singh, A.K., Patil, N.G. and Chaudhary, S.K. 2000. Management of irrigated agro-ecosystem. In. Natural Resource Management for Agricultural Productions. Ed. Yadav, J.S.P. and Singh, G.B. pp. 775-870.

Geographical Research: Issues & Solutions

Sustainable Development – Tasks Ahead

Kausalya Ramachandran

Senior Scientist (GIS Application), CRIDA

(ICAR) Santoshnagar, Hyderabad-500059

S. Padmaja

Professor, Dept.of Geography

Osmania University, Hyderabad-500007

ABSTRACT

The phrase Sustainable Development is the most commonly used term in every media and mode of communication. But the term is difficult to define because of which there is little agreement on its definition. However, it is the only rational way for Societal development in any society. Hence it's inclusion as an essential component of the educational curriculum of schools and colleges, is necessary. The paper under takes to briefly describe how we may start the process of sustainable development.

Introduction

Although the term Sustainable Development occupied center stage since it was first used in the brundtland Report in 1987, famously called "Our Common Future", it had caught the imagination of several people even before that. The term came in vogue since 1979 in one way or the other. Initially, its usage was meant to resolve the problems facing the world then, viz., over-population, pollution and energy crisis. Later its connotation meant to convey solutions for economic development. In 1982 with the entry of IUCN, the purport of natural resources conservation came about. The present definitions have a moralistic overtone. The idea that our generation is merely a trustee of natural resources of the world for the future generations has been the crux of present-day definitions of sustainable development. Many events have taken place since the notable Rio Earth Summit and Johannesburg Conference intertwined by the meetings at Doha, Marrkesh, Kyoto and Seattle However despite these many meetings and numerous initiatives there seems to be little common in our agenda for the future.

Although everyone seems to understand the purport of the term sustainable development as they can readily identify unsustainable activities undertaken by humankind, there are differences in opinion and strategies which are to be adopted in order to achieve it as noted by Murcott, (1997).This is the problem which augurs grave danger as, while vacillating we are constantly growing in great numbers and indulging in scores of unsustainable activities all at the same time through out the world (Hardi &Zdan, 1997). This state of inertia is generally detrimental and there is an urgency to turn proactive. In order to achieve sustainable development, we require to cease all unsustainable activities that directly or indirectly degrade the natural resource base namely soil cover, forest cover, number of water bodies, air quality etc.Only proper and scientific management of natural resources would ensure that our common future is secure (Swaminathan, 2000).This is easily said than done. But native intelligence of human beings has shown repeatedly that determined steps are possible if we are suitably motivated.

Present situation, unfortunately, is quite grave in many of the developing and poor countries in Asia, Africa and Latin America which face severe problems of land degradation and environmental pollution owing to increasing population and added pressure on community resource base. Obviously, allowing nature to fend for itself is not an available course; the other sensible option would he to restrict human interaction with nature. This may sound paradoxical in India but surely with 35 million persons to be added by 2020 to a present population of 1060 million at a growth rate of 1.9 percent per annum, it may well be the only option available. In the past 50 years in Asia and Africa, human beings have increased in greater numbers (over 3.8 billion in entire Asia and 855 million in the continent of Africa) when compared to Central and Latin America (approximately144 and 364 million respectively.) and elsewhere in the world. Given this ground swell "humanity must learn to live within the carrying capacity of the earth and there is no other rational option in either - the short or longer term". Unless we use the resources of the earth sustainably and prudently, we deny ourselves and our subsequent generations a future. We must adopt life styles and development paths that respect and work within nature's limits. We can do this without rejecting the many benefits that modern technology has brought, provided that technology itself works within those limits (World Bank, 2002; Krishnan. 2003).

Evidently, human civilization is at risk because of the way we live today. We are misusing natural resource and severely overexploiting the earth's natural resources in the various ecosystems. World population may double in 60 yrs. but the earth will be unable to support everyone unless there is less waste and extravagance and a more open and equitable alliance between the rich and poor. Islands of effluences cannot sustain in a sea of destitution. Even if we were to start living sustainably from this moment onwards, the likelihood of ensuring a satisfactory life for all is remote, unless the present rates of population growth are drastically reduced (UNCED. 1992). The solution and strategy to this irresolute situation seems simple and straightforward.

We must adopt a two-pronged approach: one, which secures a widespread and deeply held commitment to a new ethic of life - the ethic for sustainable living, which would strive to translate its principles into practice and another which would seek to integrate conservation with development. "The guiding principle would be conservation, in order to keep our actions within the earth's capacity and development, to enable people everywhere to enjoy long, healthy and fulfilling lives.

If this sounds an ideal way to exist, would it not be difficult to achieve? The answer is, it may be difficult but not impossible. We are exponentially increasing in number while overusing our resources and as both these aspects are at the root of the present crisis, the solution may lie in reconciling to these two issues. The question of how and where to start is uppermost in every body's mind and an attempt has been made here to address that. Although we as a nation, with a diverse resource base aim at achieving good environmental health, fertile soil for growing food and fodder, generating high income from exportable cash crops, besides development in manufacturing, tertiary and service sectors, there are numerous obstacles in achieving these goals. Much has been said, discussed, debated and written about them over the last three decades in various forum, but we seem to have made little headway. Studies have also been conducted to analyze why we have failed or succeeded so little in our efforts. It is however, time to call a halt to this dawdling and seriously take concrete measures in the following direction:

Reduction in population growth

India with a population of 1 billion persons in 2000 would increase to 1.4 billion by 2025 making it the most populated country in the world. With only 2.3% of land resources and dwindling 1.7% of world's forest stock, it would indeed be a Herculean task to achieve sustainable development. With an annual population growth rate of 1.9 % when compared to world average of 1.4 and of China at 0.9. We are in for a severe problem of over-population. Only four other nations in the world, namely - Pakistan, Bangladesh, Nigeria and Indonesia, have higher growth rate (Bossel, 1998).

Impact of climatic vagaries could be minimized

Over 296 million ha metre rainfall or 74% of all precipitation is received in the form of South-west monsoon in India, which forms the basis of agricultural activities, more so in case of dry land. This is, by no means a meager quantity but for the fact that, this is received during a few thunderstorms due to which most of it is lost as overland flows. Viable solution to this problem would be to preserve and enhance storage capacity of existing surface water bodies by desilting, dredging and removing encroachments from tank bed and it's catchment. All natural depressions must be earmarked for water collection, as human beings cannot create such large storage facilities. These natural storages would ideally provide opportunity time for percolation of water to recharge ground water table that is rapidly falling owing to overexploitation with impunity. (CGIAR, 1998; Samra & Eswaran, 2000).

Ethnic diversity not a challenge

Presently ethnic diversity is considered an impediment to the process of achieving sustainable development. On the other hand, diversity is an opportunity that brings regional variety. Indigenous technical know-how (ITK) of the diverse ethnic groups could be tapped to find solutions for local problems.

Removal of malnutrition and poor health

Demographic and social landscape crumble as market economics crowds out humane concerns. Poverty remains a serious problem with strong links to poor health. Over 40 % of India's population is suffering from malnutrition and women and children from among landless agricultural labor, urban slum dwellers and remote tribal communities are more Vulnerable. A weak nation cannot fend for itself and hence this scourge must be wiped out.

Halting industrial pollution -

Industrial pollution is a major concern, as it contaminates every aspect of environment, be it air, soil, surface water body and groundwater resource which would be impossible to redeem. Mishap like the Bhopal Gas tragedy still haunts many. Industrial air pollution in the form of suspended particulate mailer (SPM) and emissions from thermal power generating units, textile and chemical units, is rising uncontrollably. Industrial waste from petrochemical, pharmaceutical, pesticides, paint and dye. petroleum, fertilizer, asbestos, caustic soda, inorganic chemical and general engineering industries is being dumped indiscriminately on public land or in municipal solid waste sites without proper treatment. In a study carried out by us for the Ministry of Environment and Forestry, most of the Industrial Development Area (IDA) lack proper roads, drainage channels, storm waterways and waste disposal sites or facility. This leads to contamination of streams and ground water aquifers. In Andhra Pradesh. Patancheru IDA in Medak district near Hyderabad presents a disturbing scene (Kausalya Ramachandran, 2001).

Forest conservation and management

Although Land Utilization Statistics indicate a marginal increase in forest land to 14.5 % of our total geographical area of 329 million ha. it is way below the ideal of 22 % for any country. Another issue linked with forest management is that of faulty policy formulation and implementation viz., Joint Forest Management (JFM) or encouragement to industrial houses for developing forests. While JFM is restricting the entry of women folk and ethnic groups to collect firewood from forest however ludicrous it may sound, the fact of the matter remains, that women rarely hurt forest stands- Another flaw with the present forest policy is the alienation of tribal people from their natural habitat. These steps would impede sustainable development. Although capital from industrial houses to establish forest stands would be welcome, allowing them to grow tree species that are useful as raw material alone would play havoc with the ecological balance of natural stands (Katyal et al. 1996; Singh et al 2000).

Ensuring adequate water supply and sanitation facility

India is a thirsty nation despite recurring good monsoon over a large area. It is indeed an irony that people along the major rivers in the country depend on the civic water supply from deep tube wells and surface water storages for their water supply rather than the flowing water which for centuries, quenched the thirst of millions. The recent mooting of linking of rivers project is facing many skeptics who feel that pollution and not water would be redistributed. Rivers are organic beings and cannot be cut and joined like pipelines without precipitating an ecological crisis of unknown dimension (Kausalya Ramachandran, 2001, 2002).

Protection of cropland from salinity and water-logging

Making available a lot of water free or at low cost to farmers and land users have adversely affected prime cropland to the tune of over 20 million ha in the states of Haryana. Punjab, UP. Rajasthan and MP. Agricultural land must be treated as sacrosanct and those defiling it must be booked under the Environment Protection Act that rightfully provides for making the polluter pay (Singh et al. 2000).

Checking loss of soil nutrients

Overland water flow is the primary cause of soil erosion and loss of soil nutrients both major and minor nutrients causing low levels of agricultural yield forcing farmers to use unwarranted quantities of chemical fertilizers which leach and contaminate surface water bodies and ground water aquifer causing soil salinity on one hand and debt - trap for farmers, on the other (Doran et al.. 1996; Samra & Eswaran. 2000).

Halting ground water depletion and penalizing contamination

Over 16% of administration blocks in India have over-utilized their ground water resource and negligible effort has been done to increase water harvesting and recharge. Simultaneously, heavy leaching of pollutants especially from N_2 fertilizers has been observed in the past three decades that needs to be checked (Kerr et al. 2002; 001, 2001; Hanumantha Rao. 2001, 2003).

Reduce indoor air pollution

In rural India, fuel wood cause smoke, leading to respiratory and vision related problems. In urban areas the level of SPM has increased drastically owing to industrial and vehicular traffic especially in metros, according to a report published by Centre for Science & Environment. In Mumbai, for instance, higher levels of SO_2, has resulted in marked increase in respiratory and cardiac problems in people. Vehicular emissions in the form of unburnt hydrocarbon is on the rise in most of Indian metros and towns.

Obviously sustainable development cannot be achieved by mere slogans. We must resort to concrete action immediately in order to stop the slide. Following steps would help us to get a grip on the situation:

Effective implementation of the law

Clean air and water, conservation of species and tribal-forest management are a few issues that require effective implementation of law. Although as a nation, we are good at drawing policies; their implementation is often seriously flawed. In case we desire to change the scenario and halt degradation, we must ensure effective implementation of the law.

Effective population control

India must face the fact that by 2018 we would be the largest nation on earth. If the benefits of economic development have to reach the poorest of poor, we must gear ourselves to take drastic measures to control population growth rate. No nation can aspire to be a developed nation with over 50 million people living in slums, suffering malnutrition and often going hungry. Sustainable development would remain a chimera as long as there are poor who are forced to source the nature around them for food and shelter.

Reform in Food Policy and access to food by all

Food grain production has reached over 200 million tonnes and there is a buffer stock of 40 million tonnes that is spilling - over in the FCI godowns and railway yards of Punjab. However, we still hear reports of causalities from Kalahandi, Mayurbhanj and Mehaboobnagar where people are too poor to access this food resource. Steps must be taken to ensure that every Indian has access to adequate nutrition and malnutrition must be eradicated as a weak foundation cannot make a strong nation (Swaminathan. 2001: Singh & Strickland. 1993).

Ensuring Universal Education

At present access to universal education in India is merely a slogan. This is evident from the number of children we often see aimlessly moving on the road sides and in the market places or bending over arduous menial jobs for long hours every day to eke a living. The rate of school dropouts is significant and some innovative methods other than free books or mid-day meals must be innovatively adopted to keep the children in school where their thoughts could be attuned to the issue of sustainable development as an ignorant majority could easily nullify the gains made by an enlightened few. Also there is an urgent need to open newer avenues of vocation and employment for the educated youth before their disillusionment with development process. What is required at this moment is not setting up of more engineering colleges when many young engineers are jobless, but by initiating community centric programs that would open job opportunities for both the skilled and semi-skilled youth force. Sustainable development is possible only when development process ensures participation of all people who are left out from the main stream or sidelined by the process of development.

Ensuring optimum utilization of land resource

Land use must be based on land capability and suitability more popularly called its carrying capacity. This would conserve resources and check degradation that is at the heart of the problem of unsustainable development. Evidently as is often said, time is not on our side. If we have to look forward to a "common future for all", we as a nation, must first strive to have a common agenda for the benefit of all which would leave little room for doubt as to "Who will feed India?"

Reference

Bossels, H. (1998): Earth at a Crossroads: Paths to a Sustainable Future. Cambridge Univ Press, UK, 388 pp.

CGIAR (1998): Mobilizing Science for Global Food Security. Everybody complains about climate...What can agricultural science and the CGIAR do about it? Mid-Term Meeting 1998, May 25 to 29, Brasilia. Brazil

Doran, J.W. Sarrantonio, **M.** & **M.A. Liebeg** (1996): Soil health and sustainability. *Advances in Agronomy.* Vol. 56. pp 2-55.

Government of India (2001): A Report of the Working Group On Watershed Development, Rain fed Farming and Natural Resource Management for The Tenth Five Year Plan. TFYP WORKING GROUP

Sr. No. 15/2001. Government of India. Planning Commission. September. 2001

Hanumantha Rao, C. H. (2000): Watershed development in India: Recent experience and emerging issues. *Economic and Political Weekly.* 35 (45): 3943- 3947.

Hanumantha Rao, C, H. (2002): Sustainable use of water for irrigation in Indian Agriculture *Economic And Political Weekly,* 3 7(18) 1742-1745

Hardi, Peter & Terrence John Zdan (1997): Sustainable development - Evaluation. The International Institute for Sustainable Development, 161 Portage Avenue East - 6th Floor, Winnipeg, Manitoba, R3B **OY.ISBN** 1-895536-07-3

Katyal, J.C., Kausalya Ramachandran, M.Narayan Reddy & C.A. Rama Rao (1996): Indian Agriculture - Profile of land resources, crop performances and prospects. Proc. South Asia Regional Workshop on Regional land cover changes, sustainable agriculture and their interactions with global change. COS 1 ED. ICSU & UNESCO. 16-19 Dec. 1996. Madras.

Kausalya Ramachandran (2001): Assessment of Contamination of Natural Resource using GIS and Remote Sensing-a study on Hyderabad Region. GIS India. 10(1)8-13.

Kausalya Ramachandran (2001): Using GIS for Watershed Development to augment Water Harvesting in Semi-arid Telangana, AP. India. GIS India Vol.11 (5) 15-20.

Kausalya Ramachandran (2002): Land Use Planning for sustainable agriculture in rainfed regions in India. *Proc.* LUCC-ISPRS Workshop 2002. 2 December 2002. Hyderabad.

Kerr, John, Pangare, G. & Vasudha Lokur Fanfare (2002): Watershed Development Projects In India an Evaluation. Research Report 127, December 2002. IFPRI

Krishnan, R. (2003): *Swaialdhara'*. More empty promises? *Economic and Politically Weekly* 38 (10) 937-939.

Murcott, Susan (1997): -Appendix-A: Definitions of Sustainable Development. Comprehensive Data Sets of Sustainability Definitions, Principles. Criteria and Indicators of Sustainable Development. AAAS Annual Conference, 1IASA "Sustainability Indicators Symposium." Seattle. WA. 16 Feb 1997.

Samra, J.S. & Hari Eswaran (2000): Challenges in ecosystem management in a watershed context in Asia. *In* Lal, R. (2000): (Ed) Integrated Watershed Management in the Global Ecosystem. Soil and Water Conservation Society. CRC Press. Washington DC', pp 19-33.

Singli, Naresh. C. & Richard S. Strickland (1993): Sustainability. Poverty and Policy Adjustment: From legacy to Vision. Decision-makers summary findings and recommendations of the International Conference on Sustainable Development, Poverty Eradication and Macro/Micro Policy Adjustment; 2-4 December 1993

Singh, H. P., Venkateshwarlu. B., Vittal, K.P.R. & Kausalya Ramachandran (2000); Resource

Optimization in Rainfed Agro-ecosystem. Proc. International Conference on Resource Conservation & Management. 16-17"' Feb. 2000. New Delhi. Pp 669-774.

Swami Nathan, M.S. (2000): Inaugural Address. *Inter. Conf. On Managing Natural Resources for Sustainable Agricultural Production in the 21" (.century.* Fob 14-18 2000. New **Delhi.**

Swami Nathan, M.S. (200 f): Now for the evergreen revolution: Professor, a pioneer of India's green revolution, calls for a new approach to world farming. *For A Change.* August-Sept. 2001.

UNCED (1992): Agenda 21: Program of Action for Sustainable Development. UN Conf. on Environment and Development (UNCED). UN, NY. 294 pp.

World Bank (2002); the World Bank - Environmental Indicators - Environmental Economics and Indicators, <www.worldbank.org>

World Bank (2002): World Development Indicators. The World Bank. Washington DC, USA. 405 pp.

Land Transformation Processes and Issues for Geographical Research –

Scope of Remote sensing and GIS Applications

G.P. Obi Reddy, A.K. Maji, S.Thayalan and V. Ramamurthy

MBSS & LUP, Nagpur

ABSTRACT

Land transformation due to various driving forces effects biodiversity, soil quality, rainfall pattern, surface and sub-surface water availability and radiation budgets, trace gas emissions and other bio-physical processes that, cumulatively, affect global climate and biosphere. There is increasing scientific and public concern over the condition of large-scale ecological systems, including landscapes, watersheds and eco-regions or bio-geographical provinces. The concern about cumulative impacts of human activities on natural resources are particularly focused on changes in land use and environment. In the present study an attempt has been made to analyze the some of the land transformation processes of Vidarbha region in Maharashtra. The region is characterized by tropical/semi arid and sub-humid monsoon type of climate. The eastern part of the region experiences highest rainfall with an average of 1735 mm, in contrast to the western part of the region, which is receiving 588 mm rainfall. The dominant geological formations are older Precambrian, Cuddapah and Vindhyans, and Gondwana Lameta Group, Deccan Trap and Recent Alluvium deposits. The analysis of land use types in the region reveals that forest area in the region is showing decreasing trend from 28.48 to 27.68 per cent area of the Total Geographical Area (TGA) of 9.74 m ha. The area under barren/uncultivable lands was slightly increased from 2.00 to 2.35 per cent of the TGA. The area under non-agricultural use was increased significantly from 4.36 to 6.15 per cent of the TGA. Similarly crop coverage changes in total cereals, pulses, oilseeds and cotton crops have been analyzed. Some of the issues for geographical research in the areas of land use/land cover changes, their processes and impacts on environment have been identified for detailed analysis. The authors emphasized the scope of remotely sensed data and GIS applications to analyse various land transformation processes in association with climate, soil and socio-economic parameters towards development of strategies and plans for sustainable and efficient land use plans for the defined eco-systems in the Vidarbha region of Maharashtra.

Introduction

The ever increasing population pressure, over exploitation of natural resources and other developmental activities are the main driving forces for land transformation, deforestation, shrinking of surface and subsurface water resources, which are causing adverse impacts on overall environment of different regions of the globe. There is increasing scientific and public concern on the condition of large-scale ecological systems, including landscapes, watersheds and eco-regions or bio-geographical provinces and cumulative impacts of human activities on natural resources particularly changes in land use and environment. Land cover is defined as the layer of soils and biomass, including natural vegetation, crops and human structures that cover the land surface. Land use refers to the purposes for which humans exploit the land cover (Fresco, 1994). Land cover change is the complete replacement of one cover type by another, while land-use changes also include the modification of land-cover types, e.g., intensification of agricultural use, without changing its overall classification (Turner II *et al.,* 1993). Land-use changes, mostly driven by human activities, result in global environmental change (Riebsame *et al.,* 1994; Vitousek *et al.,* 1997 and Houghton, 1994). The individual activities leading to land-use changes meet locally defined needs and goals, but aggregated they have an impact on the regional and global environment (Turner II, 1994 and Ojima *et al.,* 1994). There is an increasing need to precisely describe and classify land uses and land covers in order to define sustainable land use systems that are best suited for the defined ecosystems. In order to use land optimally, it is not only necessary to have the information on existing land use/land cover but also the capability to monitor the dynamics of land use resulting out of changing demands of increasing population (Rao *et al.,* 1996).

Spatial distribution of land use/ land cover information and their changes are desirable for the planning, management and monitoring programmes at local, regional and national levels. These information not only provide a better understanding of land utilization aspects but also play a vital role in the formulation of policies and program required for developmental planning. High-resolution remote sensing data provides valuable information on location, spatial distribution and extent of land use/land cover. With increase in the spatial, spectral and temporal resolutions of the instruments, the satellite data becomes effective source for accurate land use/land cover mapping and changes with time. A number of research works have been carried out by using various methodologies and algorithms to derive land cover and changes information form different sets of remote sensing data (Gutham and Narain, 1983; Singh, 1989; Lambin, 1996; Green *et al.,* 1994; Rao, 1990 and Roy *et al.,* 1991). GIS techniques having the capability of data storage, retrieval and analysis can play an important role in land use systems studies and development. Recent advances in remote sensing and GIS technologies have emerged to meet ever-increasing demand for more precise and timely information on land use/land cover. In this direction, an attempt has been made to analyze some of the land transformation processes and issues for geographical research, which can be effectively analyzed using the latest technologies like remote sensing and GIS applications to understand the various land transformation processes in association with other parameters using the suitable models to suggest best suited cropping systems in the defined eco-systems of Vidarbha region of Maharashtra.

General Description of Study Area

Vidarbha region lies between 18^0 45' and 21^0 45' North latitude and 76^0 00' and 81^0 00' East longitudes in the eastern part of Maharashtra state with an area of 9.74 m hectares. The region is spreading in 11 districts and accounting for about 31.6 per cent of the total area of state. The region is characterized by tropical/semi arid and sub-humid monsoon type of climate. The annual rainfall of the region varies from 588 mm to 1735 mm. The mean annual maximum and minimum temperatures vary from 30.9^0C to 34.0^0C and 20.3^0C to 21.3^0C respectively. The geological formations expressed in order to antiquity are older Precambrian, Cuddapah and Vindhyans, and Gondwana Lameta Group, Deccan Trap and Recent Alluvium deposits. The region comprises of mainly Satpura ranges, Purna valley, Ajanta hills, Plateaus, Plains and Valleys. The Wainganga, Warda and Penganaga, the tributaries of Godavari, are the main rivers drain the region.

Rainfall Pattern

The analysis of average rainfall for the selected stations for the period of 30 years (1970 to 2000) of Vidarbha region shows a clear declining trend of rainfall pattern from east to west side of the region. The Dhanora station in Gadchiroli district is receiving highest rainfall in the eastern side of region with an average of 1735 mm, in contrast to Shergaon station in Buldana district of the western region, which is receiving 588 mm rainfall only (Fig.1). The rainfall pattern is the dominant factor in determining the cropping pattern, production and productivity levels of the region. The decrease trend of average rainfall from east to west has a lot of implications on available moisture for cultivated crops, groundwater availability and stress on existing forest resources. The land transformation processes in association with the rainfall pattern need to be analyzed for assessing the impact of land transformation on rainfall pattern and vice versa.

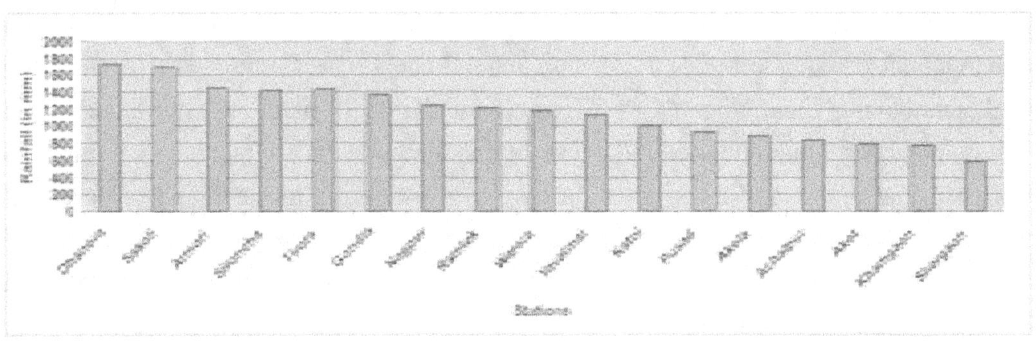

Fig. 1 Average rainfall pattern of the selected stations in the Vidarbha region

Agro-Ecological Regions of Vidarbha

The Vidarbha region is basically classified into three distinct agro-ecological regions. The western part comprising of Deccan plateau, comes under hot semi-arid ecoregion with shallow and medium (with inclusion of deep) black soils and length of growing period (LGP) ranging from 90-150 days in a year. Rainfed cotton- pigeon pea, irrigated cotton and orange orchards are the main cropping systems. The northern part consists of central high lands (Malwa, Bundelkhand and Eastern Satpura plateau) under hot (dry) sub-humid ecoregion with black and red soils and length of growing period (LGP) ranging from 150-180 days in a year. Irrigated paddy and rainfed cotton are the main cropping systems. The southern part comprising of eastern (Chhotanagpur) plateau and Eastern Ghats comes under hot (moist) sub-humid ecoregion with red and lateritic soils and length of growing period (LGP) ranging from 150-180 to 210 days in a year. Irrigated paddy and rainfed soybean- pigeon pea are the main cropping systems.

Spatial Variability of Resources

Vidarbha region is characterized by distinct spatial variability in resources distribution. The slope analysis of the region shows that nearly 61 per cent is under very gentle slopes (0-1%). The gentle, moderate and moderately steep slopes are occupied mostly in southwestern, eastern and northern parts of the region with an area of nearly 28.00, 7.72 and 2.75 per cent of the TGA respectively. The major landforms identified in the region are valleys, ridges, rolling plains, undulating terrain and level plains, which accounts for 24.00, 19.00,15.00, 21.00 and11.00 per cent respectively. The valleys are mostly occupied in northwestern part, ridges in the eastern side and rolling plains and undulating terrains in the central part of the region. The soil depth in the region shows that the deep soils accounts for nearly 62.00 per cent of the area. The very shallow and shallow soils are occupied in southwestern and northern part of the region with an area of nearly 18.00 and 10.00 per cent respectively.

Analysis of Land Transformation processes in selected cropping systems

Land utilization pattern

The analysis for the period from 1970-71 to 2000-01 reveals that the forest area in the Vidarbha region is showing decreasing trend from 28.48 to 27.68 percent area of the Total Geographical Area (TGA) (Fig.2 and Table 1). The barren/uncultivable area is slightly increasing from 2.00 to 2.35 percent of the TGA. The area under non-agricultural use is increasing significantly from 4.36 to 6.15 per cent of the TGA. The area under cultural wastelands, pasture and grazing lands is showing the declining trend. However, the net sown area in the region is remaining same. Significantly, area under sown more than once and grass cultivated area are increasing. The study of land utilization pattern in association with land and water resources, population distribution and consumption pattern helps to delineate surplus and deficit zones for resource planning on sustainable basis.

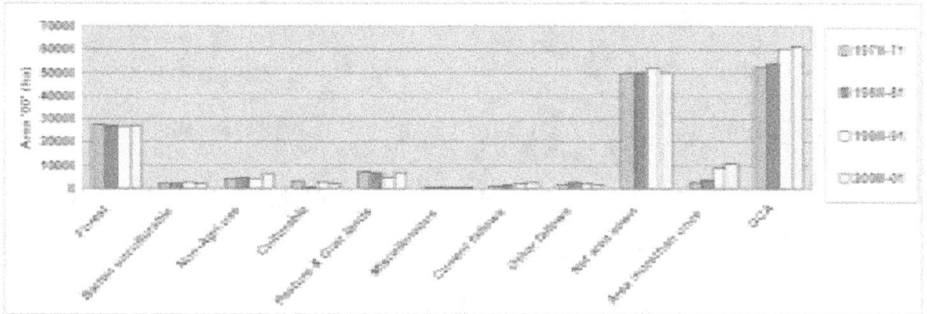

Fig. 2 Land utilization pattern in the Vidarbha region

Table 1 Area under major land utilization types of Vidarbha region

S.No	Land Use Type	1970-71	1980-81	1990-91	2000-01
1	Forest	27630 (28.36)*	27126 (27.84)	26890 (27.29)	26856 (27.56)
2	Barren/Uncultivable	1942 (1.99)	1836 (1.88)	2671 (2.74)	2288 (2.35)
3	Non-Agricultural use	4236 (4.35)	4691 (4.81)	4375 (4.50)	5970 (6.13)
4	Culturable waste	3033 (3.11)	730 (0.75)	2351 (2.41)	2099 (2.15)
5	Pastures & Grazing lands	7291 (7.50)	6955 (7.14)	4436 (4.55)	6856 (7.04)
6	Miscellaneous	671 (0.69)	592 (0.61)	713 (0.73)	647 (0.66)
7	Current fallows	1280 (1.31)	1634 (1.68)	2192 (2.25)	2711 (2.79)
8	Other fallows	1339 (1.38)	2513 (2.59)	2180 (2.24)	1544 (1.59)
9	Net area sown	49811 (51.35)	50029 (51.57)	51732 (53.33)	50003 (51.58)
10	Area sown more than once	2611 (2.69)	3738 (3.85)	8576 (8.84)	11021 (11.36)
11	Gross Cultivable Area	52422 (54.04)	53767 (55.42)	60300 (62.16)	61054 (62.94)

The values in the parenthesis indicates the percentage of TGA.

Land use changes in cereal crops

In Vidarbha region, cereals are cultivated predominantly in Bhandara, Gondia Chandrapur, Gadchiroli, Nagpur, Akola, Amaravti, Yawatmal and Buldana districts. The analysis of area under total cereals cultivation from 1960-61 to 2000-01 shows decreasing trend from 2.35 m ha to 1.66 m ha (Fig.3 and Table 2). The data indicate that significant decrease in area under total cereals in Nagpur, Chandrapur, Yavatmal and Buldana districts.

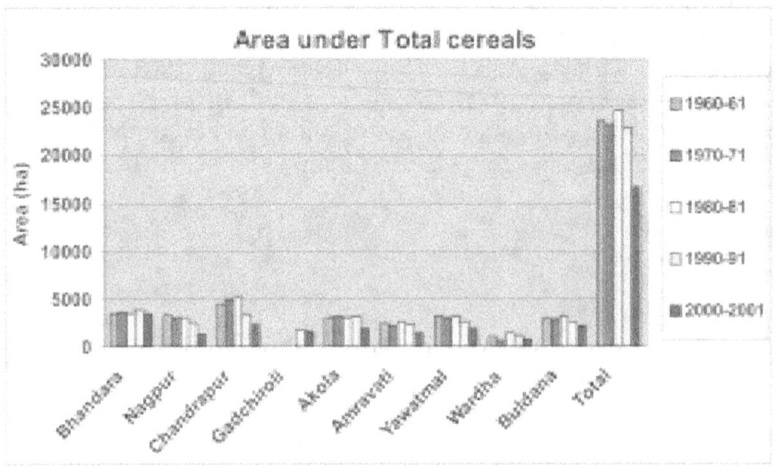

Fig. 3 Area under total cereals crops in the Vidarbha region

Table 2 Area under total cereal crops in Vidarbha region

District	1960-61	1970-71	1980-81	1990-91	2000-2001
Bhandara & Gondia	3386	3583	3365	3815	3434
Nagpur	3299	3002	3047	2358	1280
Chandrapur	4444	4886	5082	3283	2294
Gadchiroli	0	0	0	1745	1558
Akola & Washim	3027	3076	2978	3156	1864
Amravati	2441	2126	2575	2333	1473
Yavatmal	3092	2979	3156	2541	1876
Wardha	935	566	1373	1096	675
Buldana	2943	2900	3173	2532	2157
Total	23567	23118	24749	22859	16611

However, in Bhandara and Gondia districts, the area under total cereal crops cultivation increased to a considerable extent. Besides the land use changes in cereals the soil characteristics, water availability and the driving forces need to be analyzed for sustainable cereals production in the region.

Land use changes in Pulse crops

The analysis of area under pulse crops in the Vidarbha region showed significant growth from 6.71 to 13.63 per cent in 40 years period (Fig. 4 and Table 3). The analysis also revealed that area under pulses was steadily increased in the semi arid western parts of the region. However, in the sub-humid region and particularly in the Bhandara, Chandrapur and Gadchiroli districts the area under pulses are decreasing.

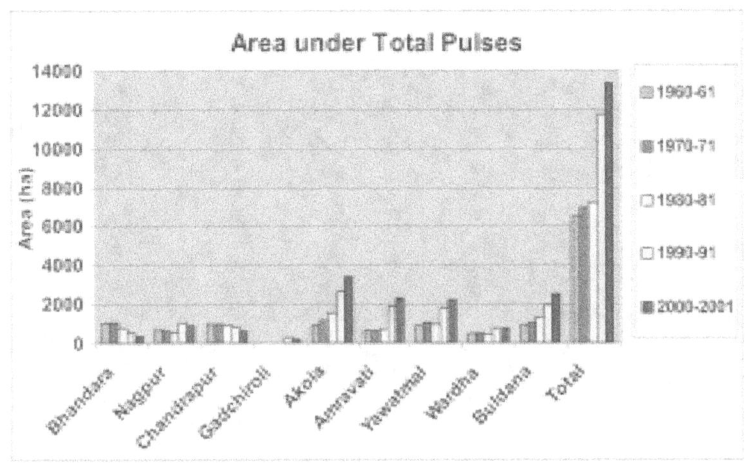

Fig. 4 Area under total pulses crops in the Vidarbha region

Table 3 Area under total pulse crops (1960-61 to 2000-01) in Vidarbha region

District	1960-61	1970-71	1980-81	1990-91	2000-2001
Bhandara & Gondia	1066	1022	753	556	381
Nagpur	652	648	573	1017	887
Chandrapur	982	964	917	832	647
Gadchiroli	0	0	0	308	220
Akola & Washim	895	1152	1501	2606	3411
Amravati	616	618	717	1905	2313
Yavatmal	907	1005	1001	1810	2218
Wardha	509	486	415	739	729
Buldana	893	1060	1338	1959	2471
Total	6540	6955	7215	11732	13277
Area in (%)	6.71	7.14	7.41	12.04	13.63

The changing rainfall pattern and market forces are the deciding factors for the changes in pulses cultivation in the region. The rainfall pattern, soil characteristics and the market forces need to be analyzed for sustainable pulses production in the region.

Land use changes in food grain crops

The analysis of area under total food grains in the region shows slight increasing trend from 1960-61 to 1990-91 and, thereafter, it is showing a decreasing trend. District-wise area under total food grains revealed the decreasing trend in Bhandara, Gondia, Nagpur, Chandrapur and Wardha districts (Fig.5 and Table 4). However, in Akola, Washim, Amravati, Yavatmal and Buldana district the area under total food grain crops is increasing steadily.

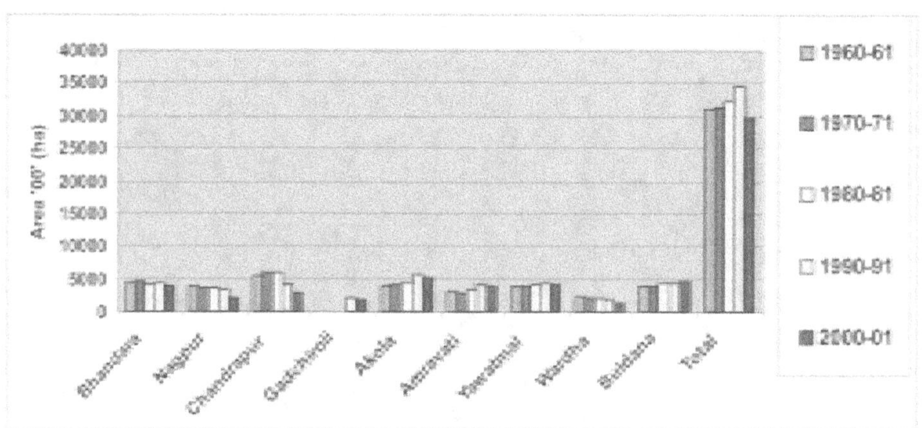

Fig. 5 Area under total food grain crops in Vidarbha region

Table 4 Area under total food grain crops (1960-61 to 2000-01) in Vidarbha region

District	1960-61	1970-71	1980-81	1990-91	2000-2001
Bhandara & Gondia	4451	4605	4118	4371	3815
Nagpur	3961	3651	3621	3375	2167
Chandrapur	5436	5850	5999	4115	2941
Gadchiroli	0	0	0	2053	1778
Akola & Washim	3921	4228	4479	5762	5275
Amravati	3058	2743	3292	4238	3786
Yavatmal	3999	3984	4157	4351	4004
Wardha	2339	2184	2149	1894	1381
Buldana	3836	3960	4511	4491	4638
Total	31001	31205	31315	34650	29775
Area in (%)	31.82	32.02	33.17	35.56	30.56

In Chandrapur and Nagpur districts area under total food grain crops was decreased significant, whereas, in Buldana district the area was increased. Besides the land transformation processes, soil suitability, food requirement and consumption patterns need to be analyzed for sustainable total food grain production in the region.

Land use changes in Oilseed crops

The analysis of area under oilseed crops (1980-81 to 2000-01) in the region reveals a significant growth from 4.24 to 11.13 per cent in the area (Fig.6 and Table 5).

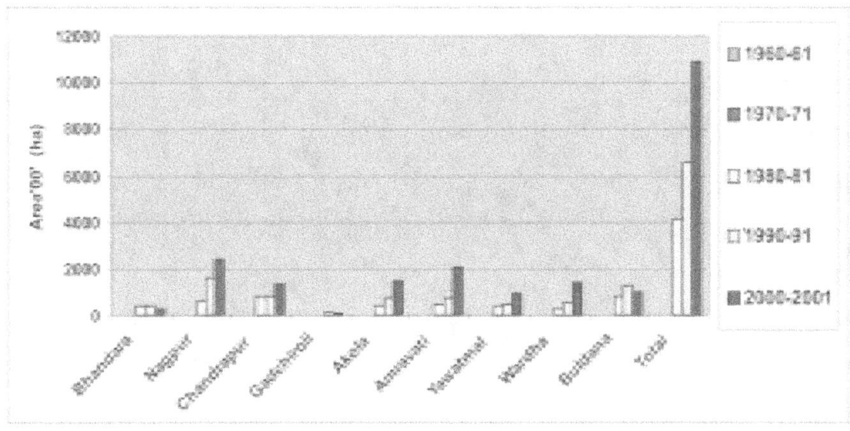

Fig. 6 Area under total oilseed crops in the Vidarbha region

Table 5 Area under Oilseeds in Vidarbha region

District	1980-81	1990-91	2000-2001
Bhandara & Gondia	398	400	235
Nagpur	619	1560	2414
Chandrapur	787	790	1347
Gadchiroli	0	113	65
Akola & Washim	418	700	1439
Amravati	442	732	2045
Yavatmal	397	463	921
Wardha	254	560	1418
Buldana	820	1237	963
Total	4135	6555	10847
Area in (%)	4.24	6.73	11.13

District wise analysis shows that the area under oil seeds cultivation was increased in Nagpur, Akola and Washim, Amravati, Wardha districts significantly, whereas marginal growth was noticed in Chandrapur, Yavatmal, and Buldana districts. However, the area under oilseed crop in Bhandara, Gondia and Gadchiroli districts decreased.

Land use changes in Cotton crop

Cotton is the most important commercial crop in the Vidarbha region. The area analysis under cotton cultivation reveals a marginal growth from 14.76 to 17.62 per cent till 1990-91 and later it showed the declining trend (Fig.7 and Table 6).

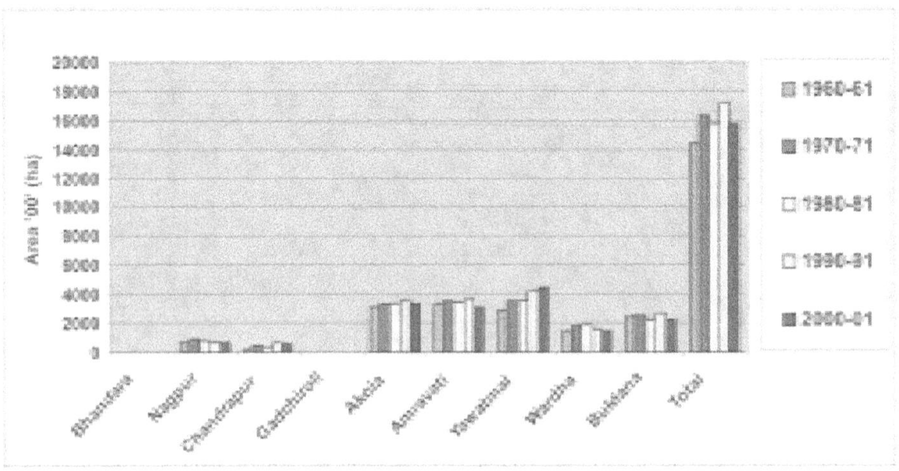

Fig. 7 Area under cotton crop in the Vidarbha region

Table 6 Area under cotton in Vidarbha region

	1960-61	1970-71	1980-81	1990-91	2000-2001
Bhandara & Gondia	0	0	0	0	0
Nagpur	700	930	795	681	711
Chandrapur	252	413	381	679	554
Gadchiroli	0	0	0	3	0
Akola & Washim	3151	3375	3378	3617	3335
Amravati	3380	3614	3507	3730	3091
Yavatmal	2938	3551	3563	4240	4477
Wardha	1504	1833	1916	1552	1404
Buldana	2452	2565	2225	2667	2198
Total	14377	16281	15765	17169	15770
Area in (%)	14.76	16.71	16.18	17.62	16.18

District wise analysis shows significant increase in area under cotton cultivation in Yavatmal, Akola and Washim districts, whereas, decline in area was observed in Amravati, Wardha and Buldana districts. The area under cotton cultivation in Amravati and Buldana districts was marginally decreased. The production levels in association with soil and climatic parameters and market facilities need to be analyzed for sustainable cotton production in the region.

Yield Gap Analysis

The analysis of total production of cereals, pulses, food grains and oilseeds in the Vidarbha region shows the marginal growth in total production of cereals, pulses and food grains from 1960-61 to 2000-01. However, significant growth was noticed under total production of oilseeds (Fig.8 and Table.7).

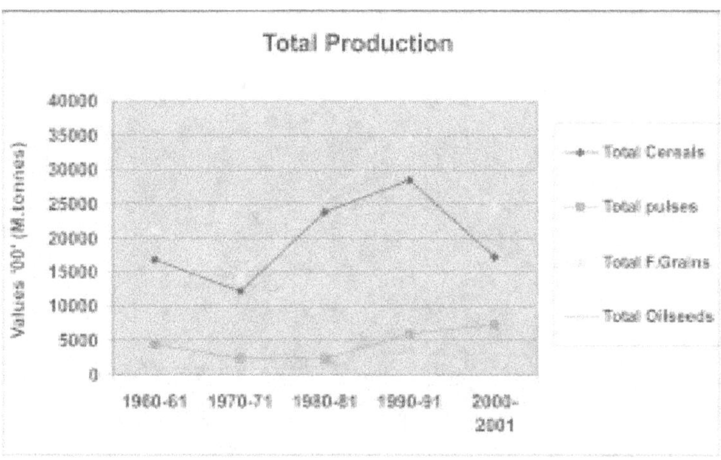

Fig. 8 Total production of major crops in the Vidarbha region

Table 7 Production levels of Major crops in the Vidarbha region (mt. tones)

Major crops	1960-61	1970-71	1980-81	1990-91	2000-2001
Total cereals	16879	12191	23693	28430	17243
Total pulses	4392	2351	2370	6072	7256
Total food grains	21271	14542	26063	34503	24499
Total oilseeds	912	683	1093	3541	9317

The productivity levels of principal crops in the region showed that significant improvement in paddy, wheat and jowar crops. Interestingly, the productivity level of cotton is almost stagnant (Fig.9 and Table.8). Probably, this might be the cause of concern for farmers besides spiraling input costs in cotton cultivation in the region. The temporal and spatial variations in production and productivity levels of major crops can be effectively studied using GIS based analysis in association with soil characteristics, water availability, food grains requirement and market facilities for sustainable production in the region.

Table 8 Productivity levels of major crops in the Vidarbha region (Kg/ha)

Major crops	1960-61	1970-71	1980-81	1990-91	2000-2001
Paddy	696	871	1101	977	867
Wheat	410	452	755	944	1019
Jowar	301	224	571	817	816
Cotton	101	30	67	142	115

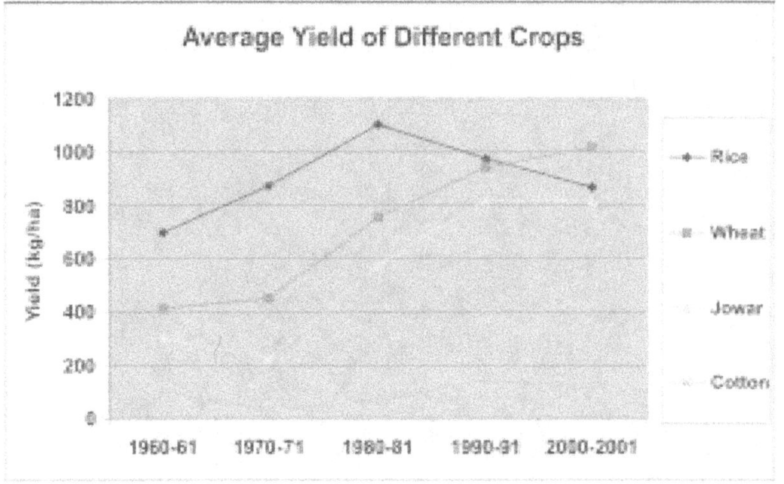

Fig. 9 Total productivity of major crops in the Vidarbha region

Land Transformation and some of the issues for Geographical research

- Analysis of land use changes and driving forces
- Land transformation and land degradation.
- Land transformation and surface and sub-surface water resources.
- Land transformation and status of wastelands.
- Analysis of cropping systems and food consumption patterns
- Land transformation and bio-mass and Organic carbon status.
- Land transformation and Livestock management.
- Land transformation and yield gap analysis.
- Prediction of near future land use changes with the help of dynamic modeling.
- Sectorial allocation of land for different uses.
- Simulation of scenarios of land-use at different levels.

- Analysis of land transformation processes and environmental risks.

- Analysis impacts of land transformation processes on biodiversity.

- Analysis of different land use systems and carrying capacity.

- Assessment of landscape patterns, composition and dynamics.

Geographical Research and Strategies

Macro-level Analysis

In macro level analysis, the spatial heterogeneity and patterns of land use/land cover have been the main focus to understand the land transformation processes and biophysical systems of the environment. The temporal analysis of remotely sensed data and analytical capabilities of GIS can be effectively used in macro-level analysis of various land transformation processes. The approach tries to unravel the processes that have caused land use to change based on observed patterns of land use over a region (De Koning *et al.* 1998; Mertens *et al.* 1997 and Liu *et al.* 1993). A macro approach appears to be a very useful to identify problems for geographical research to start with at macro-level (district as a unit) research, it can be used as lens or filter to focus on problems that require micro scale explanations. By working high up the cone of resolution, where there is little detail, it is possible to determine what aspects of the problem are relatively important. Aggregate analysis of structure and process are almost mandatory in this approach to identify the bounds of a complex system and subdivide it into more tractable components.

Micro-level (Watershed) analysis

Many a times it is difficult to analyze the various land transformation and associated processes over a region. The systematically selected micro-level studies can be carried out to understand the various processes in a given region. In micro-level analysis obviously watershed approach is the well accepted to study the various resources and their processes. High resolution satellite data and GIS based analysis will be of immense help in micro-level approach to study land use/land cover resources and their processes to assess their potential and problems and needs of the people in preparation of action plans to make it pragmatic and acceptable to the local community on long term basis.

Scope of remote sensing and GIS to address the issues of land transformation

Remote sensing and GIS techniques are being effectively used in recent times as tools to gather information about the earth's resources more accurately and quickly than conventional methods (Colwel, 1978 and Karale *et al.,* 1988). Analysis of satellite data for land use/land cover studies and GIS based analysis in conjunction with collateral data, facilitate effective evaluation of various land transformation processes. Some of the key areas have been identified to assess the land transformation processes and their impact on resources and environment.

Analysis of Land use changes and driving forces

Time series satellite data will be of immense help in the field of land use/ land cover mapping and analysis of their changes. One of the major advantages of remote sensing systems is their capability for repetitive coverage, which is necessary for change detection studies at global and regional scales. Detection of changes in the land use/ land cover involves time series satellite data. The multi-scale analysis of the driving forces of land-use changes is based on the analysis of spatial patterns of actual land-use. Except for areas with minimal human influence, the patterns reflect the result of a long history of land-use change and contain, therefore, valuable information about the relation between land use and its driving factors. Because, it is assumed that the relations between land-use and driving factors are extremely complex due to scale dependencies, interconnections and feedbacks, which need to be studied thoroughly at various levels e.g., landscape or regional or watershed level to find out the causes and their level of influence on land transformation and environment of the region.

Land transformation and land degradation

Land degradation is one major problem in the hilly and undulating terrain resulting in low productivity of agricultural land. Land transformation processes invariably affect the land quality on a long term basis in the processes to achieve maximum benefits from each land parcel. In conjunction with field surveys, remote sensing and GIS applications can help to study the spatial variability in land transformation and its degree of influence on land degradation towards generating site-specific action plans to suggest suitable mechanical and biological conservation measures to minimize the top soil loss on sustainable basis.

Land transformation and surface and sub-surface water resources

Deforestation and over exploitation of surface and sub-surface water resources leads to regional environmental changes. Detailed land transformation processes including rainfall, changing cropping patterns, forest loss and silting of surface tanks needs to be studied with the help of available remotely sensed data and GIS technologies for effective mitigation of droughts, floods and harvest the run-off water optimally. The analysis of relevant thematic database in GIS helps to develop water conservation strategies. This involves increasing the recharge from precipitation with the creation of sub-surface and surface storages with the help of various water conservation techniques like the construction of masonry weirs, excavation tanks, composite dams etc.

Land transformation and status of wastelands

Land transformation processes are closely related with status of wastelands. Lands that have one or more limitations make them generally unsuitable for the cultivation of agricultural crops and limit their use largely to pasture, forest, wildlife and recreation. The denuded forestlands and wastelands as they have a great potential for producing fodder, fuel, fiber, minor fruits and low quality timber. To achieve this, it is necessary to study spatio-temporal data with help of satellite data and GIS based techniques to adopt suitable conservation measures supplemented with proper afforestation techniques, horticultural practices grassland development etc.

Analysis of cropping systems and food consumption patterns
In view of changes in food consumption patterns, the different cropping systems including the enterprises like dairy, horticulture and vegetables should be studied. The spatio-temporal variations in changes under area under total cereals, pulses, oilseeds and food crops need to be analyzed using the high resolution satellite data for their correlation between the population distribution, consumption pattern and environment. The land transformation processes in association with soil suitability and water requirement will also be analyzed for sustainable total food grain production in the region.

Land transformation and bio-mass and Organic carbon status
Forest cover greatly influences the bio-mass levels and soil organic carbon. Vegetation moderates the local climate, prevents floods, regulates stream flow, sustains off-season discharge and protects the soil organic carbon. The study of spatio-temporal bio-mass levels with the help of time series satellite and demand of the region will be of immense help to generate action plans to protect the bio-mass and soil organic carbon on sustainable basis.

Land transformation and Livestock management
Land transformation processes effects pasture and fodder availability, grazing lands, stalk feeding and fodder supplies, grazing pressure on land, milk production, other livestock development like sheep, goats, poultry, fisheries, development of small scale/cottage industries for milk products etc. Hence, the information on the various processes of land transformation is essential to protect and improve the livestock of the region.

Land transformation and yield gap analysis
Changing cropping patterns and yields levels in association with needs on the region need to analyzed using the latest remote sensing and GIS techniques to asses the carrying capacity of the region. The analysis of yield gaps and relevant thematic database in GIS helps to potential and problematic areas to suggest suitable cropping pattern on sustainable basis.

Prediction of near future land use changes with the help of dynamic modeling
The changes in land use/ land cover due to natural and human activities can be observed using current and archived remotely sensed data. Land use/ land cover change is critically linked to natural and human influences on environment. With the availability of multi-sensor satellite data at very high spatial, spectral and temporal resolutions, it is now possible to prepare up-to-date and accurate land use/ land cover map in less time, at lower cost and with better accuracy. GIS capabilities helps in great extent to simulate spatially explicit geographical pattern of land use changes. The spatial resolution of the simulations is dependent on the extent of the study area and the resolution of data available for that study area.

Sectorial allocation of land resources for different land uses

Sectorial allocation of land for different uses based on the dynamic simulation of competition between different land use types. Competitive advantage is based on the 'local' and 'regional' suitability of the location and the national level demand for land use type related products (e.g., food demand or demand for residential area). The 'local' and 'regional' suitability for the different land use types are determined by quantified relations between land use and a large number of explanatory factors derived in the multi-scale analysis described above.

Simulation of scenarios of land-use at different levels

At different operational levels the land use scenarios need to be simulate on the basis of developments in consumption patterns, demographic characteristics, land use policies and export volumes. Different promotions/restrictions towards the allocation of land use change can be implemented, e.g., the protection of nature reserves or land allocation restrictions in areas susceptible to land degradation.

Analysis of land transformation processes and environmental risks

GIS facilitate to develop models to identify fragile environmental units, which are potentially at risk of exposure to human-induced and natural environmental stresses, which can be closely monitor to manage resources on sustainable basis.

Analysis impacts of land transformation processes on biodiversity

The impacts of land transformation on biodiversity, water and radiation budgets, trace gas emissions and other processes that, cumulatively, affect global climate and biosphere, which need to be studied at regional scales. Greenhouse emission inventories and reduction objectives need to include the foreseen changes in emission source strengths as a consequence of land-use change, rates and patterns of land-use change need to be understood to design appropriate biodiversity management.

Analysis of different land use systems and carrying capacity

The information on kind and aerial extension of different land uses are needed in the analysis of biomass estimation, carrying capacity, productivity and environmental process for sustainable production and environmental protection.

Assessment of landscape patterns, composition and dynamics

Multi-scale monitoring and assessment approaches are necessary to address the wide range of geo-environmental issues, particularly those related to landscape dynamics. Application of new concepts related to ecosystem management requires the evaluation of landscape pattern and composition at multiple spatial and temporal scales. To protect ecological balance of the region on a sustainable basis, it is necessary to look into the various physical, ecological and human induced degradation processes.

CONCLUSION

The analysis of land use/land cover changes with help of remotely sensed data and GIS will be of immense help to understand various land transformation processes to analyse the causes and consequences and their impact on environment for judicious utilization of resources to maintain ecological balance of the region. Using the GIS techniques, the anthropogenic pressure zones can be also characterized through analysis of relevant biophysical resources and socio-economic parameters. It is essential that the land and water resources in these zones to be managed properly so as to reduce pressure on the adjacent ecosystem. The remote sensing and GIS applications in analysis of land transformation can be effectively used for conservation of local ecosystems, simulate spatially explicit geographical pattern of land use changes, developments of scenarios of land-use at different levels, environmental risks, analysis of food consumption patterns, biomass levels, carrying capacity, productivity and environmental risks to develop strategies for utilization of available resources to achieve sustainable production without disturbing the ecological balance of the region.

ACKNOWLEDGEMENT

The authors are grateful to Indian Council of Agricultural Research (ICAR), New Delhi for providing financial support for Adhoc Scheme (AP-Cess Fund Project) on Vidarbha Region. The authors are grateful to Dr. K.S. Gajbhiye, Director, NBSS&LUP, Nagpur for his keen interest in execution of the work. The contributions of Mr. Sunil Meshram (Tech. Officer), Mr. Indal K. Ramteke (SRF) and Mrs. Nidhi Sharma (SRF) in data collection and analysis and Mrs. Rohini Watekar in word processing are duly acknowledged.

REFERENCES

Colwel, R.N. (1978). Remote Sensing as an Aid to the Inventory and Management of Natural Resources. Can Surveyor, 32: 183-203.

De Koning, G.H.J, Veldkamp, A., Fresco, L.O. (1998). Land use in Ecuador: a statistical analysis at different aggregation levels. Agriculture, Ecosystems and Environment 70: 321-247.

Fresco, L.O. (1994). Imaginable futures, a contribution to thinking about land use planning. In:Fresco LO, Stroosnijder L, Bouma J, van Keulen H editors. The future of the land: Mobilising and Integrating Knowledge for Land Use Options. John Wiley and Sons, Chichester. p 1-8.

Green, K., Kempka, D and Lackey, L., (1994). Using remote sensing to detect and monitor land cover and land use changes, Photogrametric Engineering and Remote Sensing, Vol.60, 331-337.

Gutham, N.C. and Narain, L.R.A. (1983). Landsat MSS data for land use and land cover mapping, A case study for Andhra pradesh, Photonirvachak, J. Indian Soc. Of Remote sensing, 11(3), 15-27.

Houghton, R.A. (1994). The worldwide extent of land-use change. BioScience 44: 305-313.

Karale, R.L., Saini, K.M. and Narula, K.K. (1988). Mapping and monitoring of ravines using remotely sensed data. J. Soil Water Cons. India, 32(1,2): 75.

Lambin, E.F., (1996). Change detection at multiple scales: Seasonal and annual variations in Landscape variables, Photogrametric Engineering and Remote Sensing, Vol.62, 931-938.

Liu, D.S, Iverson, L.R, Brown, S. (1993). Rates and patterns of deforestation in the Philippines: application of geographic information system analysis. Forest Ecology and Management 57: 1-16.

Mertens, B., Lambin, E.F. (1997). Spatial modelling of deforestation in Southern Cameroon. Spatial dis-aggregation of diverse deforestation processes. Applied Geography 17: 143-162.

Ojima, D.S., Galvin, K.A., Turner, II B.L. (1994). The global impact of land-use change. BioScience 44: 300-304.

Rao, D.P., Gantam, N.C., Nagaraja, R. and Ram Mohan P. (1996). IRS-1C applications in land use mapping and planning.

Rao, U.R. (1990). Space technology and forest management with specific relevance to developing nations. Proc. Of the 41st IAF Congress. Dresen, Germany Publication & Public relations Unit., ISRO Hq. Bangalore, pp.1-10.

Riebsame, W.E, Meyer, W.B., Turner II, B.L. (1994). Modeling land use and cover as part of global environmental change. Climatic Change 28: 45-64.

Roy, P.S., Ranganath, B.K., Diwakar, P.G, Vohra, J.P.S., Bhan, S.K., Singh, I.J. and Pandian, V.C. (1991). Tropical forest type mapping and monitoring using remote sensing, Int. J. of Remote Sensing, 12(II): 2205-2225.

Singh, A. (1989). Digital change detection techniques using remote sensing data, Int. J. of Remote Sensing, vol.10, 989-1003.

Turner II, B. L, Ross, R. H., and Skole, D. L. (1993). Relating land use and global land cover change. IGDP report no. 24; HDP report no. 5.

Turner II, B.L. (1994). Local faces, global flows: the role of land use and land cover in global environmental change. Land degradation and rehabilitation 5: 71-78.

Vitousek, P.M., Mooney, H.A., Lubchenco, J, Melillo, J.M. (1997). Human Domination of Earth's Ecosystems. Science 277: 494-499

Natural Resources & Repository, Wetlands, Ground Water & Land Use

National Land Use Policies and Strategies – the Challenges Before the Current Millennium

K. V. Sundaram

Bhoovigyan Vikas Foundation, New Delhi.

In the vision of development projected in '**The UN Millennium Development Goals'** (MDG'S), land use does not figure explicitly as a goal. But it is implied in the two related MDG'S viz. **Goal 1:** Eradication of extreme poverty and hunger and **Goal 7 :** Ensuring environmental sustainability. Land use is also an important instrumentality in the realization of what we call "**Inclusive development",** whose base or bottomline is defined by economics, ecology and equity, characterised by development, empowerment and social justice. Heavy population pressure in the developing countries and the related increased competition from different types of Land use have emphasized the need for more effective Land use planning and management. Thus, rational and sustainable Land use is an issue of great concern to our country today, as we seek to achieve the Millennium Development goals as well as preserve our precious land resources for the benefit of present and future generations.

Challenges and Tasks in Land Use Planning

The maladjustment between human society and its environments is the greatest scourge that we will be facing in the new millennium. According to a Buddhist Philosopher Daisaku Ikeda: "Maladjustment, caused by both human and natural factors, such as land degradation, droughts or famines and their whole nexus of related problems, are bound to shake the foundations of both local and International stability and order". Preventing this disaster and restoring eroded and damaged lands through appropriate land use policies and strategies is a scientific and technical challenge of the first order. It presupposes the optimization of the following tasks :

- A massive increase in agricultural production both in the medium and long term to safeguard a sufficient supply of food for the growing population that we will have at the end of this millennium.

- The repairing of ecological damage to be tackled at the same time to enable transition to an *ecologically adapted land use.*

- The need for environmentally acceptable ways of thinking, acting and addressing the issues in agricultural development, not merely from a national and regional perspective, but also from an intensely local perspective, taking note of Global imperatives.
- Blending the best of modern science with traditional wisdom and the ecological prudence of local societies.

The implementation of the above tasks requires not merely the application of a great deal of science and technology, but also a very high measure of moral/ humanitarian solidarity and participatory effort among the rural folk, who are the real stakeholders working towards meeting the challenges of a shift to the paradigm of sustainable development.

The Paradigm Shift to Sustainable Development

This paradigm shift to ecologically - oriented ways of thinking and patterns of action should proceed not merely from a local or regional perspective, but also from a Global perspective, since the **'Global Village'** envisioned by Alvin Tofler in his book on **'The Third Wave'** is now a reality. The progressive condensation, as well as integration of time and space and man's tightening bondage to the environment implied in this development transition would call for new and future-oriented concepts in framing our national policies, technologies and strategies of farming for enhanced productivity and land use change for diversification of livelihood opportunities.

This conceptual shift from recuperative to sustainable development also draws our attention to the traditional systems of land use and the need for a deepened analysis of indigenous, and ecologically well-adapted farming methods prevalent in our rural societies, particularly in agriculture and animal husbandry. Incidentally, this kind of reorientation is, of late, reflected in developed western societies also, by a movement to **"Return to Nature"** and to the traditional forms of land use and soil management.

In the above contexts, the next decade must be recognized as the decade for *"Agricultural Land Use Renaissance"*, coinciding with the commitment to the UN Millennium Development Goals. A new awareness and a new consciousness must grow during this period. Spatially-oriented disciplines like Geography and Agricultural Sciences are the main stakeholders, who can bring about this desired transformation and change. This **'search for ecologically stable land use solutions'**, known as *'eco-farming'*, is already being vigorously advocated in the developing countries. It seeks to obtain ideas for new methods of land use planning and agricultural production, by drawing upon scientific developments in natural resource management, using methods like Remote Sensing and GIS applications, besides seeking to blend the modern with the traditional production methods.

In his book on "*The Pedagogy of the Oppressed*", Freire writes that both "**clinging to the 'local', while losing sight of the *'overall view'* and taking note of only a *'bird's eye view'*, while ignoring one's *'footing or local base'* are serious mistakes.**" He insists on the importance of a balanced view. In the new millennium our policies and strategies must constantly move from the local to the global and back, to meet the concrete targets set by our international commitments, particularly our commitment to Goal 8, which is to "**develop a global partnership for development**".

Planning for Relational or contextual Space – The Human Ecosystems Approach
Land use refers to the manner of optimally utilizing *'geographic space'* for sustainable development. The term **'geographic space'** refers not only to the visible earth's surface, with its various physical features, land forms and natural resource endowments, but also the relatively non-visible *'human values'* including social relations, culture, hopes, and aspirations of the local community. This is known as **"relational space"** (i.e. the natural environment and its relationship with Man, marked by anthropogenic influences). This concept of Relational or contextual space is well known to geographers and space scientists. It is space where Earth and Man interact, to create a new landscape. Let me remind you of a famous quotation from Vidal de la Blache "**A country or Region is like a medal struck in the likeness of a people**". This is a view of space, which serves as the field for human action with their differential values. It provides a framework for physico-socio-economic planning. To illustrate the differential values conferred on space by man by an example: the space in the centre of a town is high in terms of value and can be used for building multistoried flats. On the other hand, the space in the periphery of the same town would gain in terms of valuation, by developing **'Peri-Urban agriculture'**. This shows that Geographic space can be put to its appropriate uses, depending on different valuations. Also Geographic space consisting of both the natural and cultural environment is indivisible. The lesson here is that only an *integrated approach* to the planning and management of **'geographic space'**, containing both components, will ensure that land is allocated to uses, providing the greatest *sustainable benefits*. In other words, Land use policies must be seen in the wider context of the interface of **'Earth and Man'**. This **relational context** makes a lot of difference in our perceptions and the developmental thinking process.

In the works on land use planning done by many agricultural scientists, and soil scientists, I find that the land use prescriptions are mainly determined by the physical attributes, especially climate, topography and soils, as the term *'land capability'* used in this context very clearly indicates. Studies conducted by social scientists in different parts of the world have, however, tended to discredit this approach as 'oversimplistic' and observed that environmentally similar areas have been developed in quite dissimilar ways by their inhabitants. This has given rise to the concept of "***human ecosystems***", which has underlined the idea that the human attributes of an area, which include social structures, cropping patterns, ways of life and even the culture and temperament of the inhabitants, play a crucial role in the determination of land use. A concept of "human ecosystems" thus implies the notion that individuals and groups relate to their environments, through their perception or cognition of them. This underlines the importance of inscribing human values into our *'Land use analysis'*. The concept of 'human ecosystems approach', expressed here can be injected into the land use planning concept, only through *'micro level studies'*, based on participatory rural appraisal methods.

Integration of Contextualities and Capabilities

This propulsive role of relational space poses not so much a philosophical as a methodological problem for deciding the appropriate land use for development. Here one must 'integrate' the contextualities of space and the capabilities of the humans. This should result from two processes involving different methodologies viz.

(i)   The first approach is what may be called *"**Natural Resource Management through Land capability analysis**"*. This is achieved through the use of scientific analysis, involving the application of modern technologies like Remote Sensing, GIS and Mapping to understand the type and quality of the land or 'Contextual Space' for different agricultural and other uses.

(ii)   The second approach is a process of **induction**, trying to distil out the essence and common denominators in the past activities of societies, their culture and indigenous technologies, or capabilities which have helped in conserving land. This leads to **'People centred development',** which means adopting a **"human ecosystems approach"** to land use.

The imperatives and tasks in this *congruent approach* are:

(i)   Meeting the challenging needs of today and of tomorrow through Green Revolution and Evergreen Revolution.

(ii)   Economic viability and enhanced productivity through scientific and appropriate Land use policies.

(iii)   Enhancement and maintenance of the quality of environment.

(iv)   Conservation of natural resources, particularly soil and water, which form the base of agriculture.

(v)   Efficient management of natural resources - Land, Water, Energy, Forests and Biological resources.

Our Agricultural Development Scenario: Population, Land, Forest Cover and Employment

Let us now draw the broad contours of our agricultural development scenario, or situational analysis, hitting only at those high spots of relevance to land use policies.

The population of the world, which was 2.5 billion 50 years ago, is now 6 billion and is likely to cross the 8 billion mark in the next quarter of a century. In India, it has crossed one billion mark and is expected to reach 1.4 billion in the next 25 years. While the human population is increasing in geometrical proportion, the harnessable natural resources – the land and water, which form the base to produce food, fuel, and fibre, are shrinking in size and deteriorating in quality. It is not only the pressure of human population, but also the pressure of animal population, which is seriously competing for these scarce resources. Moreover, the rapidly increasing urbanization and industrialization is further increasing the demand for land and water resources and degrading the quality of environment .

About 70% of the 1006 million people in India live in the rural areas and derive their livelihoods directly or indirectly from the 329 million hectares (mha) of land resources. Small and marginal farmers dominate Indian agriculture. Of the total number of farming families, more than 78% are small and marginal cultivators with less than 2 ha of farming land and 75 million landless workers. The availability of employment in this important sector throughout the year is dismally poor. Except in about 53.5 mha of the irrigated arable land, where the conditions may be marginally better, 90 mha of rainfed arable lands hardly provide 100 days of employment in the year. The land available per capita has shrunk from 0.30 ha in the 1950's to 0.14 ha currently. This is projected to decrease further to 0.1 ha by 2020.

On the basis of the estimates by the National Remote Sensing Agency, the total forest cover in the country is about 19.27% of the geographical area. Dense forest cover is much less and estimated to be only 11.17%.

Total unemployment in India, was estimated to be about 35 million people in 2002. According to Planning Commission's vision document, while agriculture may not lead to significant contribution to employment by way of non-farm employment opportunity, the unorganized sector, including Small and Medium Enterprises (SME's), should play a central role in the country's employment strategy.

This changing scenario is leaving no option for India, except to continuously increase productivity per unit of land and per unit of water, but at the cost of environment and sustainability. The scientific management of soil and water is the key to meet these challenges, besides the population control. This is a tall order and calls for integrated strategies for population control and for manifold agriculture productivity enhancement measures. In this context, I must mention that the high yielding seeds and high input-based technology in agriculture, which ushered in Green Revolution in the seventies and enabled us to keep food production ahead of the demand, has also become a subject of debate and calls for changes like '**organic farming**'.

The Land Use Policy formulation – A highly balancing Act

The land use debate today is being conducted amidst different theoretical currents, with their different perspectives and aims. The Soil and Space scientists, for instance, look at it as a '***prescriptive problem***'. Technologists, on the other hand, look at it as a '***problem of recreating nature' through biotechnology and genetic engineering***. Ecological scientists are concerned about ***reducing damage*** to our ecosystems. Resource management specialists are concerned with '***Conservation problems***' and the allocation of the real costs of resource exploitation. Spatial analysts are concerned about achieving ***more rational patterns of economic activities***. Behavioural scientists seek ***to reduce the unnecessary constraints on people*** and to improve their ability to make effective decisions. Economists would like the problem to be considered ***within the global economic systems*** and tend to view it in a historical process linked to economic and political structures.

The point is that not only do the aims and methods of explanation differ between the different perspectives, but also in the goals for change stipulated and the means advocated to promote the desired change. This makes the formulation of land use policies for this millennium, a highly balancing act.

The National Land Use policy Outline: A critique

Since our independence, for a long period of time, land use planning remained a neglected arena of government policy; but in 1988, it found its articulation in a policy document titled "National Land use Policy Outline (NLPO) and Action Points", which was brought out by the National Land Use and Conservation Board of the Ministry of Agriculture, Government of India. Now though dated, in the context of the staggering and sweeping changes that have come about in the national and international realms, the 19 Action Points set out in this document are still valid. It only needs some additions and re-orientations in the context of our changing times. The document confines itself mainly to actions at the national level and does not stretch its arms to the regional and local levels, where land use meets its 'client' – the farmer – and the challenges of implementation. The multi-level planning framework and perspective is lacking. Further, in the context of the winds of change, blowing in the form of globalisation, liberalisation, and privatisation, and influencing our developments in a big manner, it needs additional supporting policies to satisfy the requirements of our society, within its core values and socio-economic requirements. In this sense, the NLPO is a somewhat piecemeal formulation. More than the piecemeal nature of its formulation, it is its policy ineffectiveness that is worrying. One of its recommendations about constituting Land Use Boards at the State level suffers from inherent inadequacies. They are mostly functioning in many States without adequate legislative and technical support, and as we say *"stuck in an insufficient mould"*. But thank God, some organisation is there for its namesake and as an old saying goes "in a barren landscape, even a stunted tree is welcome". In the post-modernistic era, the Land Use Boards should emerge as effective "inspirational guide-posts", performing with admirable efficiency and giving a lead to managerial action at regional and local levels.

Land Use Policy at the National Level

Land Use Policy at the National level should effectively respond to the global imperatives of sustainable development within the free market paradigm. It should address itself to Six axioms:

(i)   Reviving a reasonably sustainable growth;

(ii)   Changing the quality of growth;

(iii)   Meeting the essential needs for jobs, energy, water and sanitation;

(iv)   Conserving and enhancing the resource base;

(v)   Reorienting technology and managing risk; and

(vi)   Blending environmental ethics and economics in decision making;

The above ingredients of sustainable development are expected to ensure that global environmental protection and national economic growth would proceed hand in hand and without any major conflicts.

Cadastral Scale Mapping For Microlevel Land Use Planning: The Need For a Participatory Approach

Let me now touch on the mapping problem. Large scale maps are useful only for macro planning purposes. On the other hand, agronomists and farmers are looking for micro-variations at the field level. The environmentalists desire such data at the watershed level. The economists prefer the data at regional level for their studies. The works that have emerged so far from soil scientists have use only at the regional or country level. The maps to benefit the farmer at cadastral scale are yet to be generated. Even when higher resolution cadastral scale maps become available, we cannot dispense with interaction with the local farmers in a participatory mode. The Bhoovigyan Vikas Foundation is using the methodology of space-related Cadastral Mapping using PRA methods, advocated by Gordon Conway, George Axinn and Nancy Axinn in the projects that it has taken up. These maps will supplement the maps using remote sensing techniques and assist greatly in micro-level planning.

Towards Convergence and Synergy

It is obvious that given the range of land use policy measures that need to be implemented at the different levels – national, regional and local – and the cooperative effort needed on the part of various agents – namely the government, community based organisations and the many individual decision-takers, constituting the large private sector – only a 'synergetic approach' can bring about the optimal use of agricultural landscapes. The task of evolving an appropriate and meaningful land use planning system in the country is challenging, but certainly not beyond the collective wisdom of our people and the scientists. I end up with a quotation. Jawaharlal Nehru observe in 1948 :

"Everything else can wait, but not Agriculture".

By Agriculture, he meant *proper Agricultural Land Use.*

Perspectives in Landuse / Landcover Change With Reference to Future Scenarios in Indian Context

P. S. Roy, Kalpana Ambastha and C.S.Jha
Remote Sensing and GIS Applications Area
National Remote Sensing Agency (Dept. of Space)
Balanagar, Hyderabad, AP.

Introduction

Land cover is determined by environmental factors - soil characteristics, climate, topography, and vegetation. Lambin et al. (2003) have identified the five classes of land change - tropical deforestation, rangeland modifications, agricultural intensification and urbanization. The major causes for land use change as cited above are : i. Resource scarcity leading to an increase in the pressure of production on resources, ii. Changing opportunities created by markets, iii. Outside policy intervention, iv. Loss of adaptive capacity and increased vulnerability, and v. Changes in social organization, in resource access, and in attitudes.

Globally, land use changes are cumulatively transforming land cover at an accelerating pace (Turner et al. 1994; Houghton 1994). These changes in terrestrial ecosystems are closely linked with the issue of the sustainability of socio-economic development and in turn affect our natural capital such as climate, soils, vegetation, water resources and biodiversity (Mather and Sdasyuk 1991).

Despite improvements in land-cover characterization made possible by earth observing satellites (Loveland et al. 1999), global and regional land covers and, in particular, land uses are poorly enumerated (IPCC 2000). Concerns have been raised about the large magnitude of change and according to one of the available estimates - the global expansion of croplands since 1850 has converted some 6 million km^2 of forests/woodlands and 4.7 million km^2 of savannahs/grasslands/steppes and within these categories, respectively, 1.5 and 0.6 million km^2 of cropland has been abandoned (Ramankutty and Foley 1999). Land-cover modifications / changes in the structure of an extant cover of a short duration, such as forest succession under slash-and burn cultivation are also quite significant.

These factors interact mutually in slow or fast manner to bring about the land use changes. Increase in urban sprawls has contributed much to land cover change. Among the demographic factors the major determinants of land use are population size and density, technology, level of affluence, political structures, attitudes and values of local people. Developing countries have much faster urban population growth - an average annual growth rate of 2.3%, compared to the developed world's urban growth rate of 0.4%. According to available figures, in 1950, only 18% of the population in developing countries lived in cities, the proportion was 40% in 2000 and by 2030 the developing world is projected to be 56% urban. It has also been estimated that 60% of this urban growth in 1960-1990 was from natural increase, 40% from in-migration from rural areas and the expansion of urban boundaries. During 1974 –1999, New Delhi's population increased by 4.2 million and there was a corresponding loss of 60,000 hectares of agricultural land.

Land-use/land-cover change as linked with global environmental concerns

Land-use and land-cover changes are so pervasive that, when aggregated globally, they significantly affect key aspects of Earth System functioning. They directly impact biotic diversity worldwide (Sala et al. 2000). They contribute to local and regional climate change (Chase et al., 1999) as well as to global climate warming (Houghton et al. 1999). Land uses as deforestation, mining, urban spread are the primary source of soil degradation. By altering ecosystem services, they affect the ability of biological systems to support human needs (Vitousek et al. 1997). Such changes also determine, in part, the vulnerability of places and people to climatic, economic or socio-political perturbations (Kasperson et al. 1995).

There are a number of global environmental issues that are directly linked with the land-use/ cover changes and they include: a) Regional/global food security; b) Global climate change impacts on regional land and water use; c) Climate variability and regional vulnerability; d) Post-Kyoto carbon issues; e) Regional biodiversity loss; and f) Socially and/or environmentally fragile regions. Currently there is growing unanimity on the fact that landuse change is a major driver of global change, through its interaction with climate, ecosystem processes, biogeochemical cycles, biodiversity and even more importantly - human activities. It is needed to understand the patterns and processes of land-use change and the human responses to land use/land cover change, and integrate them into global and regional models and development of databases on land surface, biophysical processes and their drivers.

Role of modeling in the generation of future scenario and its major objectives

The main objectives for land use land change studies and modeling has been to improve understanding of, and gain new knowledge on regionally based, interactive changes between land uses and covers, especially as manifested in modeling approaches. A more specific goal is the development of improved means for projecting and back casting land uses and land covers.

The role of LULC change modeling acquires larger importance in the context of fast pace of change in the various categories of land use / land cover in a fast developing country like India. In the last 50 years the food grain production has increased from 50 million tones to 212 million tones. Despite this progress, the food grain production still depends on the rainfall, which varies considerably spatially and temporally. In last 50 years there have been 15 major droughts. The very food security of India may be at risk in future due to the threat of climate change leading to changing frequency and intensity of droughts and floods.

Various modeling efforts have shown that the forest boundaries as well as species composition may get affected in the event of likely climate change. Studies have projected a higher level of NPP in the Indian forests in the event of elevated levels of CO_2. These changes may lead to vulnerability of the forested ecosystem. The other natural ecosystems such as grasslands, mangroves and coral reefs are also likely to be affected. The increasing CO_2 levels may favour the growth of C3 however the elevated temperatures may favour the growth of C4 plants. It can be said that the changes would vary spatially for different ecosystems.

Similarly, the Indian coastal zone is very long (about 7500 km long) and fragile. The Indian coastal strip is heavily populated (the population density being 1.5 times higher than the national average). The likely climate change may further complicate the coastal problems such as erosion, flooding, submergence and deterioration of the coastal ecosystems.

Water resources in India pose most serious challenges for the long-term sustainability of the national population. India shares about 16% of the global population but has only 4% of the total water resources. The varying pattern of rainfall coupled with the drought and flood cycles make it very difficult to manage the river waters where the water levels gets completely depleted and the reverse in the monsoon season. It is apparent that a thorough understanding and prediction of future change scenario in the LULC in the Indian context is very much needed before the irreversible climate change driven by the LULC change brings the insurmountable problems.

LUCC project, a long-term international research project jointly developed under the auspices of the International Geosphere-Biosphere Programme (IGBP) and the International Human Dimensions Programme on Global Environmental Change (IHDP) addresses the following questions (Lambin et al. 1999) 1. How has land cover been changed by human use over the last 300 years? 2. What are the major human causes of land-cover change in different geographical and historical contexts? 3. How will changes in land use affect land cover in the next 50-100 years? 4. How do immediate human and biophysical dynamics affect the sustainability of specific types of land uses? 5. How might changes in climate and global biogeochemistry affect both land use and land cover? 6. How do land uses and land covers affect the vulnerability of land-users in the face of change and how do land-cover changes in turn impinge upon and enhance vulnerable and at-risk or critical regions?

Land Use-land Cover Change Modelling

Models of land use and land cover change are powerful tools that can be used to understand and analyse the important linkages between socioeconomic processes associated with land development, agricultural activities and natural resource management strategies and the way that these changes affect the structure and function of ecosystems (Turner and Meyer 1991). The modeling of land-use change can either be empirical, involving diagnostic models based on an extrapolation of the patterns of change observed over the recent past, with a limited representation of the driving forces, or dynamic, with integrated models based on an understanding of the processes of land-use change. Diagnostic models integrate landscape variables and proximate causes of change in a data-rich spatial context. However, they can only provide short-range projections (5 to 10 years at most) due to non-stationarity in land-use change processes.

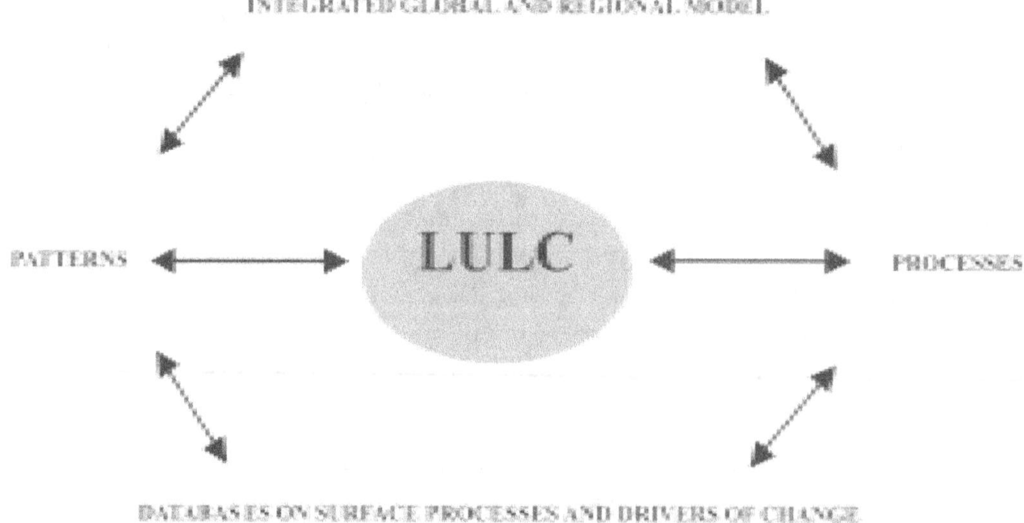

Fig. 1 The broader research themes of LUCC

(Lambin et al. 1999)

Longer-range projections require an understanding of major human causes of land-cover changes in different geographical and historical contexts, as well as an understanding of how changes in climate and global biogeochemistry affect land use and land cover.

Predictive models of landscape (inclusive of land use and Land cover) change in a human-inhabited landscape must describe the social processes that affect land use (e.g. development, agricultural production, tourism and recreation). Because land use change occurs parcel-by parcel, where the parcel is the basic unit of land ownership, information should be collected, and processes modeled, with the parcel as the basic unit. To characterize the biophysical implications of land-use change, e.g. on biodiversity, water quality, and carbon sequestration, the relationships between land use and land cover must also be quantified. Land-cover change is not restricted to parcel boundaries and sub-parcel representation of land cover is preferable. Because land use and land cover are related but not equivalent, models of landscape change should include this missing link. Additionally, by coupling land-use and land-cover change models, such a link may provide modelers of land-use change with a means to use remote sensing data to help validate land-use change models (Brown et al. 2000).

A generalized understanding of the drivers of land-use change, which can be linked to patterns of change at the regional scale, is gained through comparative analysis of these case studies of land-use dynamics.

Major issues in methodologies

Models of land use at regional to global scales include a representation of economy-environment linkages. They have to cope with such issues as heterogeneity and scales, technological innovations, policy and institutional changes, and rural-urban dynamics. Regional scenarios and assessments can be generated for projecting future land-use changes or for identifying land-use patterns with certain optimality characteristics. These models are used to evaluate the impact of policy or climate change on land use (Lambin et al. 1999).

Whereas most models were originally developed for deforestation (Kaimowitz and Angelsen 1997; Lambin 1997) and more recent efforts also address other land use conversions such as urbanization and agricultural intensification (Brown et al. 2000; Lambin 1999). Spatially explicit approaches are often based on cellular automata that simulate land use change as a function of land use in the neighborhood and a set of user-specified relations with driving factors (Wu 1998; Verburg et al. 2002). The generic methodological issues can be summarized as follows :

1.Coping with heterogeneity and scales in regional models - commodity heterogeneity, agents (social and cultural heterogeneity), landscapes (spatial heterogeneity); 2.Improving the environment - economy linkage; 3.Dealing with technological change; 4.Policies and institutions - representing the regulatory context in regional models of land-use/land-cover change.

Matters of Scales

Environmental management, and land-use planning specifically, take place at different spatial and organizational levels, often corresponding with either eco-regional or administrative units, such as the national / state or at district level. The information needed and the management decisions made are different for the different levels of analysis. At the national level it is often sufficient to identify regions that qualify as "hot-spots" of land-use change, i.e., areas that are likely to be faced with rapid land use conversions. Once these hot-spots are identified a more detailed land use change analysis is often needed at the regional level. At the regional level, the effects of land-use change on natural resources can be determined by a combination of land use change analysis and specific models to assess the impact on natural resources.

National level – Land cover changes are addressed by adopting one of the following methods: 1) Through direct observations and diagnostic models of land-use/land-cover changes, 2) Land-cover change, hot spots and critical regions - monitoring biophysical and socio-economic variables, definition of land-cover change indicators and developing Hot spot detection and alarm system, 3) Socializing the pixel -Transition probability models, Spatially explicit models, Linking patterns of land-cover change to household level data and 4) From patterns to processes- Dynamic, causal models and short-range projections, definition of risk zones and potential impacts.

Regional and Global modelling mainly need to include the following: 1) Development of a framework and tools for integrative assessments, 2) Review and comparison of past and current regional modelling studies, 3) Land-use/Land-cover change and the dynamics of interrelated systems - Rural-urban dynamics, Water issues in regional land-use/land-cover change, 4) Expanding the global food and fiber production and 5) Scenario development and assessments of critical environmental themes, 6) Scenario development in the context of a growing world population and 7) Landuse / land-cover change in global environmental assessment.

It is required to aggregate data at different spatial scales. A way to do this is to organise both the biophysical and socio-economic data in their respective hierarchies. Subsequently, these hierarchies must be compared and linked (matched) spatially. To avoid this discrepancy, matching may require the framing of artificial scales based on grid aggregations. A major disadvantage of this grid approach is that one may loose information, firstly because the minimum grid size becomes the most detailed level of analysis possible and secondly because of the borders of units which normally do not fit into one grid cell. A third disadvantage is the artificial nature of the units of analysis. However, once data are converted into grid units, similar and equally sized units can be compared without any spatial aggregation problem. Another advantage is that artificially gridded data can be aggregated into many different scales while for example data grouped in administrative boundaries can only be aggregated into a few predetermined scales.

In most cases, the models are connected to a raster-based geographic information system (GIS). The models predict which grid cells are likely to experience future land-cover change. IGBP-IHDP (Report 10) has summarized three main activities to address the LUCC change strategy as follows :

Table 1 Main Activities of LULC Project

Activity1: Land-Use Dynamics	Activity2: Land-Cover Changes	Activity3: Regional and Global Models
Comparative analysis	Direct observations and diagnostic models	Integrative assessments
1.1 Understanding land-use decisions	2.1 Land-cover change, hot-spots and critical regions	3.1 Review and comparison of past and current regional modelling studies
1.2 From process to pattern: linking local land-use decisions to regional and global processes	2.2 Socializing the pixel	3.2 Major issues in methodologies of regional land-use/land-cover change models
		3.3 Landuse / Landcover change and the dynamics of interrelated systems
1.3 Sustainability and vulnerability scenarios	2.3 From patterns to processes	3.4 Scenario development and assessments of critical environmental themes

(Lambin et al. 1999)

Types of models

Land use land cover change models can be used for different purposes. These models can be categorized according to amount of information they contain. These are Whole landscape models, Distributional landscape models, or spatial landscape models (Baker 1989).

Spatial details plays important role in these processes (White et al. 1997). Therefore spatial modeling has more relevance than other methods of modeling in research. Different approaches have been attempted in spatial modeling. To name a few models, based on approaches 1) Cellular Automata models, 2) Artificial neural network models 3) Multi Agent models 4) Statistical models 5) Fractal models.

Cellular Automata Models (CA)

The approach in this model is 'bottom to top'. The final global structure emerges from purely local interactions among the cells. CA not only offers a new way of thinking for dynamic process modeling it also provides an opportunity for testing the decision making processes. CA has natural affinity with GIS and remotely sensed data (Torrens et al. 2001). One of the most significant properties of CA is perhaps its simplicity.

Transition probabilities have been used extensively for analysis and modeling of land-use and land-cover change (Burnham 1973; Bell 1974; Turner 1987; Muller and Middleton, 1994). The approach treats state transitions as Markovian random processes that are conditional on the initial state only. However, there are primary limitations of Markov-based transition probability-based models for landuse and land-cover change analyses such as the assumption of stationarity in the transition matrix and assumption of spatial independence of transitions (Brown et al. 2000). The transition probabilities are often derived empirically from multi-temporal maps with no description of the process (Baker 1989).

Artificial Neural Network Models (ANN)

Artificial Neural network consists of simple processing elements, called neurons, and connection links operating in parallel. Two or more of the neurons can be combined in any layers. A network may contain one or more layers. Typical three kinds of layers in neural network architecture are known as input layer, hidden layer and output layer. The layer between input and output layer is called as hidden layer, because they have no direct relationships with outer real world. However, the number of hidden layers and neurons required to obtain an accurate approximation could not be unanimously decided yet. ANN model is taught by sample data taken in real area as a human brain learns, thinks and reacts against stimulus. During the training, initial weights that are assigned to interconnection links are modified repeatedly until the ANN can produce acceptable outputs that matches the original target values even though not exactly the same.

Agents based models

Multi-agent (MA) systems are designed as a collection of interacting autonomous agents, each having their own capacities and goals but related to a common environment. This interaction can involve communication, i.e., the passing of information from one agent and environment to another. An agent-based model is one in which the basic unit of activity is the agent representing actors in the situation being modeled, often at the individual level. They can be developers, individuals, state policy etc. their influence can be at different scales. Agents are autonomous in that they are capable of effective independent action, and their activity is directed towards the achievement of defined tasks or goals. They share an environment through agent communication and interaction, and they make decisions that tie behavior to the environment (White and Engelen 2000).

Agent-based models of Land-use and Land-Cover change (ABM/LUCC) combine a cellular model representing the landscape of interest with an agent-based model that represents decision-making entities. The cellular model may include a variety of spatial processes and influences relevant for LUCC. It may draw on a number of specific spatial modeling techniques, such as cellular automata, spatial diffusion models, and Markov models. The agent-based model provides for an extremely flexible representation of heterogeneous decision makers, who are potentially influenced by interactions with other agents and with their natural environment. Thus, ABM / LUCC are well suited for analysis of spatial processes, spatial interactions, and multi-scale phenomena (Parker et al. 2002)

Spatial-Statistical models

Transition probabilities have been used extensively for analysis and modeling of land-use and land-cover change (Muller 1994). Whereas, Markov transition probabilities provide a convenient analytical framework for simulating land-cover change using observed transitions, e.g., from remote sensing, alternative approaches are typically used for modeling the influence of social and economic drivers on land-use change. Traditional statistical models, e.g. Markov chain analysis, multiple regression analysis, principal component analysis, factor analysis and logistic regression, have been very successful in interpreting socio-economic activities but they needed to have spatial component within themselves so that they can used to their full potential in geography. However, integrating the spatial component does not solve the problem as time and spatial domain do not follow standard distribution like normal distribution. Therefore the sampling technique is also questioned.

Fractal based model

Fractals are spatial objects having properties of 1) self-similarity (scale independent), 2) fractional dimension. They can be formed by repeating themselves.

Accuracy of Models

New integrated and regional models, informed by empirical assessments of the patterns of land use and case studies that explain the processes underpinning such configurations of land-use and land-cover change over varying spatial and temporal scales are near-term needs for more precise land-use and land-cover change projections arising from the IPCC, the Framework Convention on Climate Change (UNFCCC), and other international bodies. A review of existing regional to global scale agricultural, grassland and forestry models for basic stock-taking and identification of the main gaps in current methodologies should be followed by creation of new modelling structures and tools capable of more completely capturing and explaining land-use and land-cover change and its main driving forces.

There is need for investigation concerning the criteria that LUCC modelers use to evaluate models and validation (Mertens and Lambin 2000). Scientists usually do not test the performance of LUCC models versus the performance of a Null model (Fielding and Bell 1997).

Predictive power of a model, then there must be a clear distinction between the procedures of calibration and validation. Calibration is "the estimation and adjustment of the model parameters and constraints to improve the agreement between model output and a data set", whereas validation is "a demonstration that a model within its domain of applicability possesses a satisfactory range of accuracy consistent with the intended application of the model" (Rykiel 1996). Separation of the calibration process from the validation process is one of the best ways to assure that the model is not over-fitted.

Pontius et al. (2004) have suggested to use a validation technique that: (a) budgets the sources of error, (b) compares the model to a Null model, (c) compares the model to a Random model, (d) performs the analysis at multiple scales. The following factors determine the model's predictability (Pontius et al. 2004).

Interpretation of scale

The model performances are basically scale dependent. As Pontius et al. (2004) discovered at resolutions less than 1 km, the Null model performed better than Geomod, which performed better than the Random model but at resolutions coarser than 1 km, both Geomod and the Random models perform better than the Null model. However, Geomod and the Random models are nearly indistinguishable beyond the 1 km resolution.

The statistical criterion

Most models use percent correct as the conceptual foundation to evaluate the agreement between maps. Some other measures include number of patches, patch size, patch density, contagion, fractal dimension, etc. However, the model must first improve its prediction of the quantity of the categories in order eventually to improve the model's prediction of patch pattern. Also, if the model predicts the categories in generally the correct neighborhood, then the pattern of patches are likely to be similar.

The bias of masking

Pontius et al. (2004) have pointed out that any masking can introduce bias into the validation assessment and to avoid it the entire study area should be used in the analysis.

Scientists must develop statistical methods to validate land use/ land cover models, because it is essential to know a model's prediction accuracy (Pontius and Schneider 2001). Relative operating characteristic (ROC) has been used as a quantitative measurement to validate a land-cover change model (Pontius and Schneider 2001).

Few important ongoing **LULC Project and models around the globe** are as follows :

1) TREES. JRC Tropical Ecosystem Environment Observations by Satellite, 2) SYPR - Land-cover and land-use change in the southern Yucatán Peninsular region, 3) IIASA-LUC. Modelling land-use and land-cover changes in Europe and northern Asia, 4) CLUE - The Conversion of Land Use and its Effects, 5) IMPEL- Integrated Model to Predict European Land Use, 6) GEOMOD2, 7)

CLUE (Conversion of Land Use) and its Effects

The Conversion of Land Use and its Effects (CLUE) modeling framework (Veldkamp and Fresco 1996, Verburg et al. 1999) was developed to simulate landuse change using empirically quantified relations between land use and its driving factors in combination with dynamic modeling. In contrast to most empirical models, it is possible to simulate multiple land-use types simultaneously through the dynamic simulation of competition between land-use types.

GEOMOD2

GEOMOD2 a successor to GIS-based model GEOMOD quantifies factors associated with land-use, and simulates the spatial pattern of land-use forward and backward in time. GEOMOD2 reads rasterized maps of land-use and other biogeophysical attributes to determine empirically the attributes of land that humans tend to use. Then GEOMOD2 uses the patterns of those biogeophysical attributes to simulate the spatial pattern of land-use change. GEOMOD2 can select locations for land-use change according to any of three decision rules based on (1) nearest neighbors, (2) stratification by political sub-region, and/or (3) the pattern of bio-geo-physical attribute.

Land Transformation Model (LTM)

Land Transformation Model (Pijanowski et al. 1996; Pijanowski et al. 2000) couples Artificial Neural Network (ANN) routines to GIS databases containing information on population growth, transportation factors, and locations of important landscape features such as rivers, lakes, recreational sites, and high-quality vantage points to forecast future land use patterns.

Case Studies From India

In the past few decades there is change in land use, because of expansion of mining areas, increment in construction of dams, industrialization, urbanization etc. to name, a few which affect the areas as an external factors. Internal changes includes shifting cultivation areas, selective logging due to human pressure on forest resources and habitat loss of wildlife due to reduction in the forests.

Studies so far conducted in India on LULC change is scattered particularly in regions like Western & Eastern Ghats, Himalayas and Northeastern states. In Western Ghats over the past centuries the changes in landscape is mainly due to plantations (tea, coffee, rubber, teak etc.) and some due to anthropogenic pressure. Menon and Bawa (1998) have estimated the rate of deforestation in the Western Ghats to be 0.57% annually during the period 1920–1990 and Prasad et al. (1998) have assessed 0.90% annual decline in natural forest cover in Kerala for the period 1961–1988. Deforestation has been particularly intensive in the southern Western Ghats, which lost a quarter of its forest cover between 1973 and 1995 (Jha et al. 2000). The data from Agastyamalai region, Western Ghats indicating a five-fold increase in forest loss from the periods 1920–1960 to 1960–1990, also suggest that the rates may be increasing (Ramesh 1997). Pontius and Batchu (2003) studied land use change pattern and also calibrated and validated the disturbance in the Western Ghats of India (1920 – 1990). Giriraj (2005) studied vegetation and land cover changes in KMTR, Tirunelveli where 121 km^2 of intact evergreen habitat were converted into secondary / semi-evergreen forest in a time span of 30 yrs as a direct function of socio-economic changes, land use patterns with biogeophysical characteristics. Menon et al. (2001) have predicted the land cover changes i.e. the areas most susceptible to future deforestation and biodiversity loss in Arunachal Pradesh using GEOMOD2. The authors have predicted a 50 % loss in the state's 1988 forest by 2001, mostly in areas with no legal protection. Based on these predictions authors have identified four categories of conservation prioritization areas. Dadhwal and Chhabra (2003) have made a detailed study of long-term historical land use/cover changes and their impacts on the agro ecosystem carbon cycle in Indo-Gangetic Plains Region (IGPR) states, estimated an increase of 435.6 Mt in crop biomass for the period 1901- 1991 leading to an increased agricultural contribution to carbon emissions. Using IPCC methodology, the estimated total 1990 CO_2 emissions from energy, industrial, agriculture, waste and land use change and forestry sectors are 585 Tg for Indian IGPR states, compared to 76.1 and 39.9 Tg for Pakistan and Bangladesh respectively. Under the ISRO/DOS Geosphere - Biosphere Programme, all the major river basins of India have been taken up for studying the changes that has taken place in the last two decades in relation to human dimensions.

Satellite remote sensing derived vegetation/ land use maps of years 1980, 1989 and 2000 (Talukdar et al. 2001) has been used to characterise the dynamics in Meghalaya (state in North eastern India). A hierarchical geospatial model was run to map the status of land dynamics in two decades. The characterization provides valuable input to conservation, as it is one of the important biodiversity hotspots.

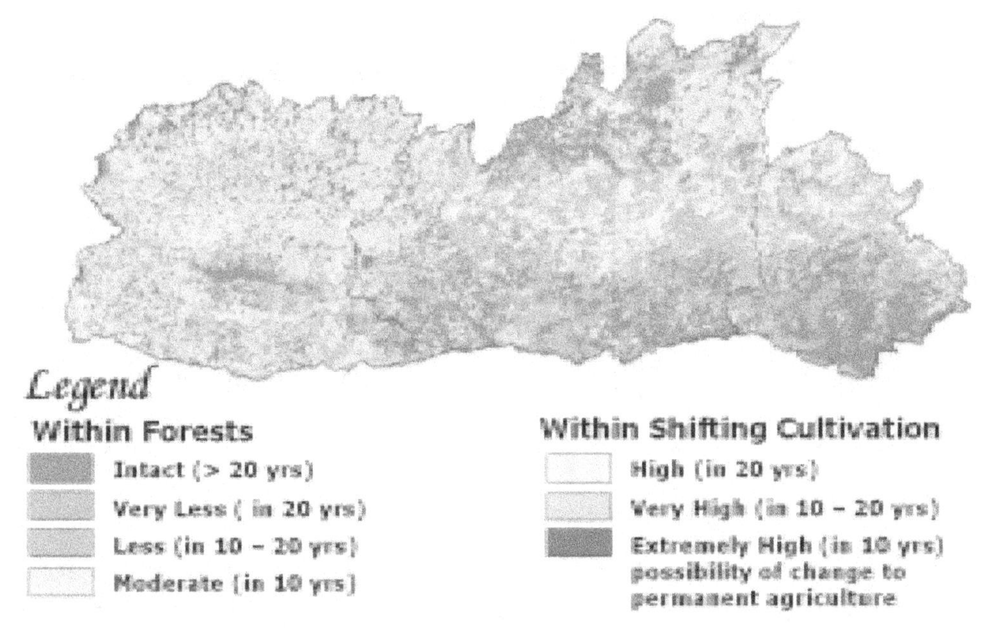

Fig. 2 Land cover dynamics in Meghalaya

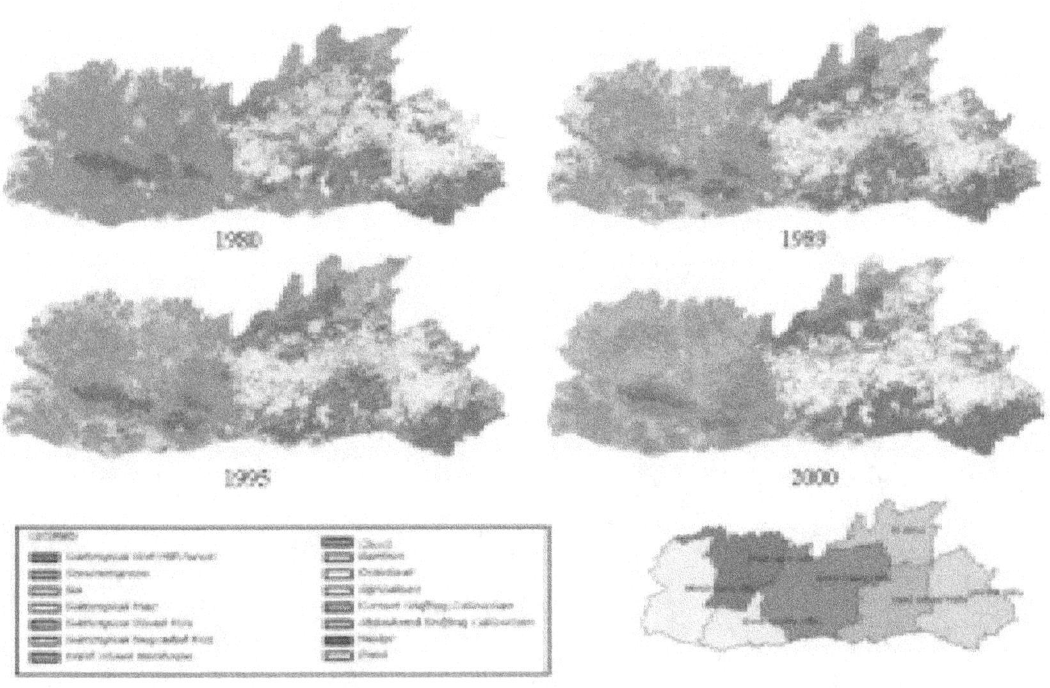

Fig. 3 Landscape Dynamics in Meghalaya (Talukdar et al. 2001)

Temporal changes in Meghalya

The comparative status of the forest cover of Meghalaya during 1980 to 1995 has been shown in table 1. The forest cover of Meghalaya has been decreased during this time period. The exact figures of the total forest cover in 1980 could not be ascertained because of the more cloud cover 769.95 sq km (3.43%). The forest cover of Meghalya has decreased during 1980-1995. The trend of forest cover shows that during 1980-89 maximum deforestation has been done. However, the negative trends during 1989-1995 shows that the deforestation process has slowed down (Fig. 3).

Table 2   Meghalaya forest cover as assessed by satellite data

Land use category	1980		1989		1995	
	Area (Sq km)	Area (%)	Area (Sq km)	Area (%)	Area (Sq km)	Area (%)
Forest	15489.28	69.06	14260.45	63.29	14157.42	63.09
Non forest	6169.77	27.51	8128.80	36.54	8247.55	36.80
Cloud Cover	769.95	3.43	39.75	0.17	24.03	0.11

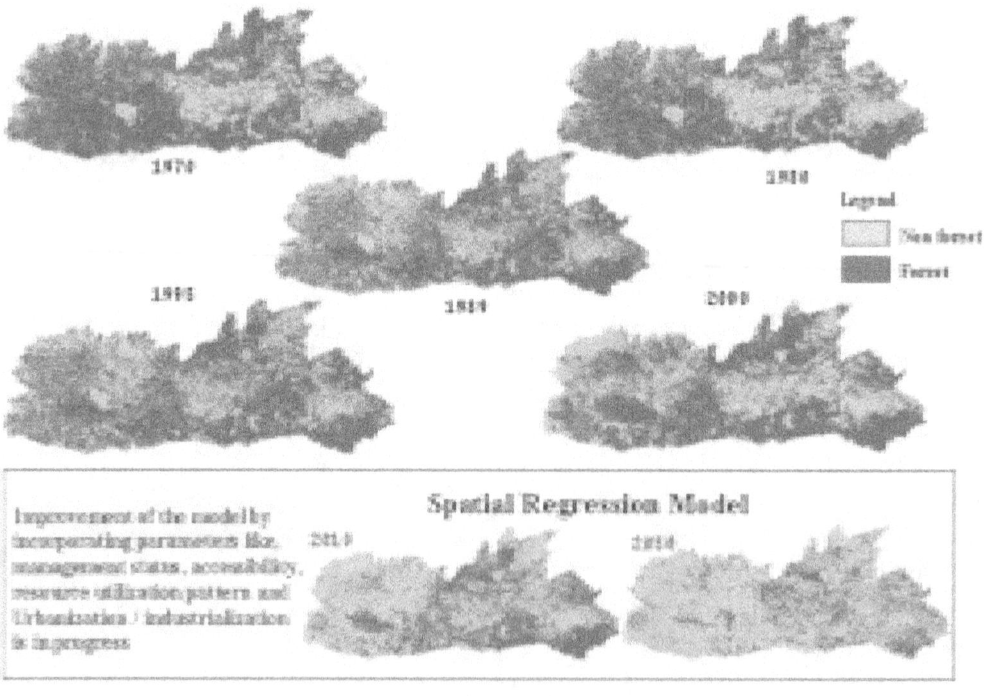

Fig. 4 Changes from 1970 to 2000 and modeled for 2050

The degradational activities viz., shifting cultivation and clear felling of forests for timber and mining has altered the natural landscape to a great extent. This has resulted in fragmentation of the landscape and loss of many endemic species of the state (Haridasan and Rao 1985). There has been significant increase in landscape variability during 1980 – 1995. The land transformations result in the alternation of natural habitats. These changes have brought in impacts like fragmentation, loss of biodiversity and degradation of sites. Spatial presentation of landscape dynamics can be used to infer disturbance regimes horizontally. Higher land cover dynamics has been observed in Garo hills showing more alternations in the landscape during the study period. For example unique plant species such as *Cycas pectinata* and *Dipteris wallichi* were lost in Meghalaya (Kataki 1983). Also species such as *Diospyros undulata*, *Nymphaea pygmaea* and *Sageretia hamosa* are still considered extinct and *luvunga scandens* is thought to locally extinct (Khan et al. 1997; Roy and Tomar 2000).

LULC Assesment in India – The DOS Experience

In India the information on LULC in the form of thematic maps, records and statistical figures are inadequate and do not provide an up to date information on the changing land use patterns and processes. Over the years, the efforts made by the various Central / State Government Departments, Institution / Organizations etc., is sporadic and often efforts are duplicated.

The Indian experience on use of satellite data for LULC analysis mainly comes from studies conducted at National Remote Sensing Agency (NRSA) in collaboration with different agencies. Realizing the need for an up to date nationwide LULC maps by several departments in the country, as a prelude, a LULC classification system (with 24 categories up to Level-II, suitable for mapping on 1:250,000 scale) was developed by NRSA, DOS, taking into consideration the existing land use classification adopted by NATMO, CAZRI, Ministry of Agriculture, Revenue Department, AIS & LUS etc. and the details obtainable from satellite imagery. The classification system provided the conceptual frame-work after discussions with nearly 40 user departments / institutions in the country and 22 fold classification system which was adopted for Nationwide LULC Analysis.

Nationwide LULC Analysis for Agro-Climatic Zone Planning

District-wise LULC analysis of all the 15 agro-climatic zones, using the 22 fold LULC classification system was completed using 1988 – 89 satellite data sets. IRS - LISS-I data of kharif (July-October) 1988 and Rabi (November– March) 1989 were used to generate details of crop land in Kharif and Rabi seasons, the area under double crop, fallow lands, different types of forest, degradation status, wasteland, waterbodies etc. NRSA along with Regional Remote Sensing Service Centres (RRSSC's), State Remote Sensing Centres and other institutions completed this task using hybrid methodology i.e., visual as well as digital methods. Out of 442 districts in the country, 274 districts are analyzed using visual techniques and remaining 168 districts by digital techniques. Planning Commission of India was the main user for this project.

National Wastelands Inventory Project (NWIP)

Until recently, no attempt had been made to prepare map showing different types of wastelands in India. The area reported by various government agencies on the extent of wastelands varies from 38 M Ha. to 175M. Ha. In 1985, NRSA/Department of Space prepared wasteland maps of all states and union territories at a 1:1 million scale. The total area of wastelands in the country during 1980 – 1982, estimated through this study was about 53.3 million ha or 16.2 per cent of the area of the country.

In 1985, National Wastelands Development Board (NWDB) was setup with the objective of rehabilitating 5 million ha of land each year for fuel wood and fodder production through a massive programme of seeding and afforestation. This programme required a very reliable database that provided details on the type, extent, location and ownership of wastelands. Confronted by varying estimates of the extent of wasteland, including the NRSA figure based on remote sensing, it became evident that the NWDB had to provide precise definitions of the various categories of wasteland. The Technical Task Force established by the NWDB proposed a classification system consisting of thirteen categories of wasteland.

Subsequently waste land mapping on 1: 50,000 scale was taken up in five phases and about 5000 wasteland maps covering the country were prepared. Methodology developed based on pilot studies were used for the identification and delineation of different types of wasteland using enlarged satellite data. Both Landsat Thematic Mapper (TM) and Indian satellite (LISS-II and LISS-III) data were used for mapping purposes. Hybrid methodology i.e. both visual and digital techniques were used to extract the wastelands thematic details. About 63.87 million ha (20.17 per cent) have been estimated as wastelands through this study.

Forest / Vegetation cover analysis

Using Landsat MSS data of 1972-75 and 1980-82 periods, NRSA (1983) carried out vegetation mapping on 1:1 million scale for the entire country, which showed a substantial decrease in forest cover. Subsequently, Forest Survey of India (FSI), Ministry of Environment and forests (MOEF) carried out mapping of India for the year 1981-83 which was published in the year 1987. Since 1987, the forest cover of the country is being assessed biennially by Forest Survey of India (FSI) using remotely sensed data. FSI has carried out 8 such surveys using satellite imagery of the periods 1981-83, 1985-87 , 1987-89, 1989-91, 1991-.93, 1993-95 and 1996-97. The total forest cover amounts to only 19.39 per cent of the geographical area of the country as per the latest available report of FSI.

Land Cover Mapping using Spot-Vegetation for South Central Asia

Under Global Land Cover (GLC) 2000, Indian Institute of Remote Sensing (IIRS), Dehradun, India has carried out a study for South Central Asian Region as part of this programme. The study has been executed with a participation of network support from countries like China, Sri Lanka, Myanmar, Thailand, Bhutan, Nepal and Bangladesh. The study has produced LULC map for South Central Asian Region using SPOT-4 VEGETATION and other ancillary information.

Biome level characterization of Indian Vegetation (IRS– WiFS Data)

Realising the potential of the IRS – WiFS datasets for regional level mapping, the assessment of phenological growth of vegetation in forest eco system has been attempted under ISRO-GBP programme by Indian Institute of Remote Sensing (IIRS), Dehradun . The climatic data with bio geographic map is used to delineate the biomes in the Indian Sub continent.

Vegetation type mapping

As part of landscape level biodiversity characterization project (DOS-DBS supported programme), vegetation type mapping of NE regions and Western Ghats on 1:250,000 scale was done using IRS-LISS-III satellite data. Central India, Eastern Ghats and East coast are being mapped on 1 : 50,000 scale using IRS P6, LISS-III satellite data

Vegetation and land use and land cover (IRS WiFS data)

Realizing the importance of LULC over a period of time, regional level mapping has been attempted in India using IRS WiFS data to bring out causes, rates, magnitude, patterns and trends in landscape changes. Earlier studies have demonstrated substantial contribution of RS and GIS techniques to map spatial distribution of important habitats

Integrated Mission Sustainable Development (IMSD)

This is one of the important projects carried out by Department of Space (DOS). It was initiated in 1987 as 'Integrated Study to Combat Drought'. Under this project different thematic maps viz., LULC, Hydrogeomorphology, Soils, Slope etc. were generated on 1: 50,000 scale and integrated to derive locale specific prescriptions called action plans for sustainable development of land and water resources. The entire work was carried out in three phases covering 175 districts in different agro-climatic zones covering about 84 million ha. or 25% of the total geographical area (NRSA,2002).

NRIS project

DOS has initiated this project in continuation to IMSD and the digital databases are being prepared for various themes. In this project the standards for database design, structure, theme content and codification were evolved. This project is being implemented in 17 states and all important natural resources including the LULC are being mapped on 1: 50,000 scale using IRS-LISS III data.

Integrated Resources Information System for Desert areas (IRIS-DA)

This is one of the recent projects carried out at NRSA (2002-2005) for Ministry of Rural Development (MRD). It covers parts of four states – Rajasthan, Karnataka, Gujarat and Haryana. In this project all thematic maps of natural resources are prepared on 1: 50,000 and the action plans for land and water resources development are generated. The LULC theme was mapped up to Level-III classes. In this project the action plans were generated using fuzzy logic and output were generated through an automatic software programmes developed.

Wetlands of India

This project is being carried out by SAC, Ahemdabad with the objective of mapping all the wetlands (like marshes, swamps, open water bodies, mangroves, tidal flats etc.) on 1: 250,000 scale for most of the states and on 1: 50,000 scale for few small states and UTs. This project was sponsored by Ministry of Environment and Forests, Government of India. Wetland delineation and mapping has been done using IRS-LISS I/II data of 1992/1993. The total wetland area has been estimated to be 7.6 M ha (excluding Paddy, Rivers and Canals).

Land Use / Land Cover inventory under NR Census

Before taking up of national level NR Census Mission, proto type districts studies were taken up to develop legends for mapping various themes, standardization of methodology including digital data base creation and generate census statistics for various natural resources. Prototype studies were done using 1999-2001 databases to understand the degree and magnitude of changes over a period of 5 years. DOS had contacted line departments under each theme to address above points and to create awareness in them.

The different themes addressed are geomorphology, soils, land degradation, LULC, wet lands, vegetation types and forest crown density, snow and glaciers. Prototype districts were selected based on diverse terrain types, physiography, agro-climatic zones etc. in the country.

Landuse / Landcover Mapping – AWIFS

Based on the pilot experience of the NR census project, this nation wide mapping project of LULC was undertaken in which National level LULC will be mapped using multi-temporal AWiFS data on 1:250,000 scale for every cropping season of Kharif, Rabi and Zaid (summer) to have level 1 (8 classes) and Level II (23 classes) classification.

Conclusion

With the ever-growing economic growth in the landuse / landcover, the developing countries have experienced significant changes and is believed to have impacted the local / regional climate directly / indirectly. This phenomenon is also directly linked to the larger issues like sustainability and food security issues and thus a thorough understanding of the growing

LULC change process, its causes / agents /drivers and predictive modeling has acquired immense importance in the recent times. The reliable estimation of various change categories at regional / global levels depend solely on the robust modeling. In India, we have been generating a large amount of geospatial databases as part of several nationwide projects and now time has come when the scientists and academia should prioritize the modeling of the LULC in the Indian context. These inputs will also strengthen the NATCOM estimates addressed to "Kyoto Protocols" and UNFCC requirements as signatories of the protocols.

References :

Baker, W. L (1989). A review of models of landscape change. Landscape Ecology 2(2): 111-133.

Brown, D. G., Duh, J. D. and Drzyzga, S. 2000. Estimating error in an analysis of forest fragmentation change using North American Landscape Characterization (NALC) Data. Remote Sensing of Environment 71, 106–117

Chase, T.N., Pielke, R.A., Kittel, T.G.F., Nemani, R.R., and Running, S.W. 1999. Simulated impacts of historical land cover changes on global climate in northern winter. Climate Dynamics 16, 93–105.

Dadhwal, VK and Chhabra, A. 2002. Landuse/landcover change in Indo- Gangetic plains: cropping pattern and agroecosystem carbon cycle. In: Abrol YP, Sangwan S and Tiwari MK (Eds) Land use: Historical Perspectives, Focus on Indo-Gangetic Plains. Allied Publishers, New Delhi, 667p

Fielding, A.H. and Bell, J.F. 1997. A review of methods for the assessment of prediction errors in conservation presence/absence models. Environ. Conserv. 24 (1), 38–49.

Haridason, K. and Rao, R.R. 1985. Forest Flora of Meghlaya, Volume 1. Bishen Singh Mahendra Pal Singh, 23-A Connaught Place, Dehra Dun.

Houghton, R.A. 1994. The worldwide extent of land-use change. Bioscience 44: 305-313.

Intergovernmental Panel on Climate Change. 1996. Climate Change 1995: Impacts, adaptations, and mitigation of climate change: Scientific-technical analyses. R.T. Watson, M.C. Zinyowera, and R.H. Moss (eds). Contribution of Working Group II to the Second Assessment Report of the Intergovernmental Panel on Climate Change. Cambridge University Press.

Jha C.S., Dutt C.B.S. and Bawa K.S. 2000. Deforestation and land use changes in Western Ghats, India. Current Science 79: 231-238.

Kaimowitz, D. and Angelsen, A. 1997. Economic models of tropical deforestation: a review. Bogor Centre for International Forestry Research (CIFOR): Indonesia.

Kasperson, J.X., R.E. Kasperson, and B.L. Turner II, (eds). 1995. Regions at Risk: Comparisons of Threatened Environments. United Nations University Press: Tokyo.

Kataki, S.K., 1983. Some rare plants in Khasi and Jaintai hills of Meghalya. In: An Assessment of Threatened Plants of India (S.K. Jain and R.R. Rao eds), Howarh: Botanical Survey of India, 149-150.

Khan, M.L., S. Menon and Bawa K.S. 1997. Effectiveness of the protected area network in biodiversity conservation, a case study of Meghalya. Biodiversity and Conservation 6: 853-868.

Lambin, E. F.1997. Modelling and monitoring land-cover change processes in tropical regions. Progress in Physical Geography 21, 375–393

Lambin, E.F., Baulies, X., Bockstael, N., Fischer, G., Krug, T., Leemans, R., Moran, E.F., Rindfuss, R.R., Sato, Y., Skole, D., Turner, B.L. II, and Vogel, C. 1999. Land-use and land-cover change (LUCC):Implementation strategy. IGBP Report No. 48, IHDP Report No. 10, Stockholm, Bonn.

Lambin, E. F, Geist H. J. and Lepers E. 2003. Dynamics of land-use and land-cover change in tropical regions, Annu. Rev. Environ. Resour. 2003. 28:205–41

Loveland, T.R., Zhu, Z., Ohlen, D.O., Brown, J.F., Reed, B.C and Yang, L.M. 1999. Analyses of the IGBP global land cover characterization process. Photogrammetric Engineering and Remote Sensing 65 (9), 1021–1032.

Mather, J.R. and Sdasyuk, G.V. 1991. Global change: geographical approaches. University of Arizona Press: Tucson.

Menon S., Pontius R.G Jr., Rose J., Khan M.L., and Bawa K.S. 2001. Identifying consevation priority areas in the tropics : a land use change modeling approach. Conservation Biology. Vol. 5, 501- 512.

Mertens, B. and Lambin, E. 2000. Land-cover-change trajectories in southern Cameroon. Ann. Assoc. Am. Geogr. 90 (3), 467–494. 85, 253–270.

Menon S. and Bawa K.S. 1998. Tropical deforestation: Reconciling disparities in estimates for India. Ambio 27: 576-577.

Muller, M. R. and Middleton, J. 1994. A Markov model of land-use change dynamics in the Niagara Region, Ontario, Canada. Landscape Ecology 9, 151–157

Parker, D.C., Berger, T. and Manson, S. M. (Eds). 2002. Agent-Based Models of Land-Use and Land-Cover Change. Report and Review of an International Workshop. LUCC Focus 1 Office, Indiana university, Belgium, vii.

Pijanowski, B.C., S.H. Gage, D.T. Long and W. C. Cooper. 2000. A Land Transformation Model: Integrating Policy, Socioeconomics and Environmental Drivers using a Geographic Information System; In Landscape Ecology: A Top Down Approach, Larry Harris and James Sanderson eds.

Pontius, R. G. Jr. and L. C. Schneider. 2001. Land-Cover Change Model Validation by an ROC Method for the Ipswich Watershed, Massachusetts, USA. Agriculture, Ecosystems and Environment 85(1-3):239–248.

Pontius Jr R.G. and Batchu K. 2003. Using the Relative Operating Characteristic to Quantify Certainty in Prediction of Location of Land Cover Change in India. Transactions in GIS 7: 467–484

Pontius R.G Jr , Huffaker D. and Denman K. 2004. Useful techniques of validation for spatially explicit land-change models. Ecological Modelling 179, 445–461.

Prasad S.N., Vijayan L., Balachandran S., Ramachandran V.S. and Verghese, C.P.A. 1998. Conservation Planning for the Western Ghats of Kerala: I. A GIS approach for location of biodiversity hotspots. Current Science 75: 211-219.

Ramankutty, N., Foley, J.A. 1999. Estimating historical changes in global land cover: Croplands from 1700 to 1992. Global Biogeochemical Cycles 13, 997–1027.

Ramesh B.R., Menon S. and Bawa K.S. 1997. A vegetation based approach to biodiversity gap analysis in the Agastyamalai region, Western Ghats, India. Ambio XXVI : 529-536.

Roy, P.S. and S. Tomar. 2000. Biodiversity characterization at landscape level using geospatial modeling technique. Biological Conservation, 95(1): 95-109.

Rykiel Jr., E.J. 1996. Testing ecological models: the meaning of validation. Ecol. Model. 90, 229–244.

Sala, O.E., Chapin, F.S., Armesto, J.J., Berlow, E., Bloomfield, J.,Dirzo, R., Huber-Sanwald, E., Huenneke, L.F., Jackson, R.B., Kinzig, A., Leemans, R., Lodge, D.M., Mooney, H.A., Oesterheld, M., Poff, N.L., Sykes, M.T., Walker, B.H., Walker,M. and Wall, D.H.2000. Biodiversity: global biodiversity scenarios for the year 2100. Science 287, 1770–1774.

Tolba, M.K. and El-Kholy, O.A. (Eds.). 1992. The World Environment 1972–1992:Two Decades of Challenge. Chapman & Hall, London.

Torrens, P. M. and D. O'Sullivan. 2001. "Editorial: Cellular automata and urban simulation: where do we go from here?" Environment and Planning B 28: 163-168.

Turner, B. L. and Meyer, W. B. 1991. Land use and land cover in global environmental change: Considerations for study. International Social Sciences Journal 130, 669–667.

Turner II, B.L., W.B. Meyer and Skole, D.L.1994. Global Land-Use/Land-Cover Change: Towards an Integrated Program of Study. Ambio 23 (1): 91-95.

Talukdar, G, M.C. Porwal, Harnam Singh, and Roy, P.S. 2001. Temporal patterns of Landscape fragmentation in Meghalaya. IGBP Report. (Communicated)

Verburg, P. H., P. Schot, M. Dijst, and Velkamp, A. 2002. Land-Use Change Modeling: Current Practice and Research Priorities. GeoJournal.

Vitousek, P.M., Mooney, H.A., Lubchenco, J. and Melillo, J.M. 1997. Human domination of earth's ecosystems. Science 277, 494–499.

White, R. and Engelen, G. 2000. "High resolution intergated modelling of the spatial dynamics of urban and regional systems." Computers, Environment and Urban Systems 24: 383-440.

White, R., G. Engelen and Uljee, I.1997. "The use of constrained cellular automata for high resolution modeling of urban land use dynamics." Environment and Planning B 24: 323-343.

Wu, F. 1998. "Simulating urban encroachment on rural land with fuzzy-logic-controlled cellular automata in a geographical information system." Journal of Environmental Management 53(4): 293-308.

Remote Sensing And GIS Application For Natural Resource Mapping

***Abhishek Srivastava, **Shweta Srivastava *Vandana Tiwari**

*Student Remote Sensing & GIS Division (University of Allahabad), **Swarnpath GIS Forum

ABSTRACT

Remote Sensing helps in providing reliable information on various aspects of natural resource. It helps in gathering information on land use, land utilization pattern in any region. Identification of wastelands and wetlands is possible by using this technique. Remote Sensing plays a very vital role in mapping of soil types, soil productivity, soil reaction, soil texture, soil degradation, and soil taxonomy, organic carbon status, fertility status etc. Extent of forest and density of forest cover can be measured by the remote sensing data. Spatio- temporal encroachment in notified forestlands can also be identified .The discrepancies in the data related to various aspect of land use can be mitigated with the help of remote sensing. In case of agriculture, remote sensing is a very important tool. With the help of remote sensing area under Rabi crops, Kharif crops, cropping intensity, crop production can be assessed and analyze. This will lead to identification of food deficit as well as surpluses areas, district wise distribution of food grains productions forecasts can also be made on the bases of yield models. On this basis the user is facilitated to find out the various levels of risk areas, identification of food deficit as well as surpluses areas, district wise distribution of food grains productions forecasts.

Studies on natural resource management have brought out the basic concept that methods of recovery and use of natural resources are closely interrelated with many human dimensions. Planning and management are facilitated by the application of spatial technology based primarily on Geographic Information System (GIS) and Remote Sensing (RS) techniques. Spatial technology is a powerful tool for conducting spatial analysis and presenting the dynamics of various spatial features.   GIS plays an important role in the assessment of natural resources, monitoring and controlling the activities of resource degradation and creating inventory of resource depletion. It helps in the formation of Environmental Information System (EIS). Development of generic decision support system helps using databases and models for resource management planning at micro, meso and at macro levels. GIS helps in identifying the protected areas from environmental point of view, in pursuing environmental risk assessment analysis, identifying high, medium, low risk areas. It also provides the grounds to find out the potential areas and location points for monitoring, day-to-day situation of environmental degradation, and remedial action to be taken to minimize the environmental hazards.

The topic "Remote Sensing for Natural Mapping" is selected due to the importance of natural resources in every sphere of life in both direct and indirect way right from the use of water to use of soil for food production for very basic need of survival. These resources are being used for this very basic need to various other needs for human. Now it is very important to study and have a very close look of these resources, which are depleting at a very fast rate and hence can be some day come to complete non-existence and thus puts on the question mark on the existence of life on earth. So a complete study, mapping of these resources as well as estimation of its abundance can tell us about its presence, its modes of utilization, its rate of degradation and various measures to conserve it for its sustainable use. Remote Sensing data (aerial & satellite) are applied for mapping and monitoring of various natural resources related with Earth's surface. Since aerial photograph and satellite image provide a wealth of detail information of a large area of Earth surface as a one time permanent record, inaccessible terrain can be surveyed fairly well. Remote Sensing has proved to be very effective means for developing and integrated GIS, which could meet the challenges of evaluating and managing natural resources. The extent and the amount of resources present, changes in resource, potential, strategy for resource protection and conservation; eco system studies, sustainable use and development practice of natural resources all can be effectively studied by integrated approach with Remote Sensing & GIS. Studies on natural resource management have brought out the basic concept that methods of recovery and use of natural resources are closely interrelated with many human dimensions. The main purpose of this study is to learn the methodology of natural resources inventory of a region and to map them on a suitable scale using Satellite Remote Sensing and Geographical Information System techniques, also to identify various natural resources and to know the methods of preparation of the various thematic maps such as Geology, Geomorphology, soils, forestry, land use, hydrology and the work of man at near the surface of the earth with in a very short time in an economic manner.

Introduction

Nature contains a plethora of resources right from the soil, water, and land to various other forms and structures like forest, animals, minerals etc., with the ever increasing demand and exploitation of resources it is mandatory to use it sustainably. The topic "Natural Resource Mapping" is selected due to the importance of natural resources in every sphere of life in both direct and indirect way right from the use of water to use of soil for food production for very basic need of survival. These resources are being used for this very basic need to various other needs for human. Now it is very important to study and have a very close look of these resources, which are depleting at a very fast rate and hence can be some day come to complete non-existence and thus puts on the question mark on the existence of life on earth. So a complete study, mapping of these resources as well as estimation of its abundance can tell us about its presence, its modes of utilization, its rate of degradation and various measures to conserve it for its sustainable use. Remote Sensing data (aerial & satellite) are applied for mapping and monitoring of various natural resources related with Earth's surface. Since aerial photograph and satellite image provide a wealth of detail information of a large area of Earth surface as a one time permanent record, inaccessible terrain can be surveyed fairly well. Remote Sensing has proved to be very effective means for developing and integrated GIS, which could meet the challenges of evaluating and managing natural resources. The extent and the amount of resources present, changes in resource, potential, strategy for resource protection and conservation; eco system studies, sustainable use and development practice of natural resources all can be effectively studied by integrated approach with Remote Sensing & GIS. Studies on natural resource management have brought out the basic concept that methods of recovery and use of natural resources are closely interrelated with many human dimensions. The main purpose of this study is to learn the methodology of natural resources inventory of a region and to map them on a suitable scale using Satellite Remote Sensing and Geographical Information System techniques, also to identify various natural resources and to know the methods of preparation of the various thematic maps such as Geology, Geomorphology, soils, forestry, land use, hydrology and the work of man at near the surface of the earth with in a very short time in an economic manner.

Study Area

As the name suggests the natural resources in abundance can be found in the natural state of earth, which are away from the artificial environment of man. Though the area also contains some human settlement, which is necessary to look on the modes of utilization of resources. The selected area is near the protected forest of Satna district and contains Barua River as its main source of water having many tributaries. So, the area contains three basic combinations of vegetation, soil & water as natural resources. This study of natural resources can provide valuable inputs for policy formulation & planning at various desired level.

Physiographical Setting

The Upper Barua River Basin lies in between the **$80^0 36$'E to $80^0 43$'E longitude and $24^0 16$'N to $24^0 25$'N latitude.** Moist sub-humid climate are mainly found in the target area. Study area which has got **340-600m** variation of contours height falls under plateau of Vindhayan Super group. The study area is drained by Barua river, a tributary of river Tones, which is further sub-divided into tributaries having Dendritic: medium to fine pattern. The study area occupies the rocks of Vindhayan Super group. The study area has undulating features with its northern part have Cuesta form of morphology, along with this the north western part have got moderately dissected lower plateau and attached with it is Mesa. Middle portion of the study area have got undulating surface mass, which contains potential water resource.

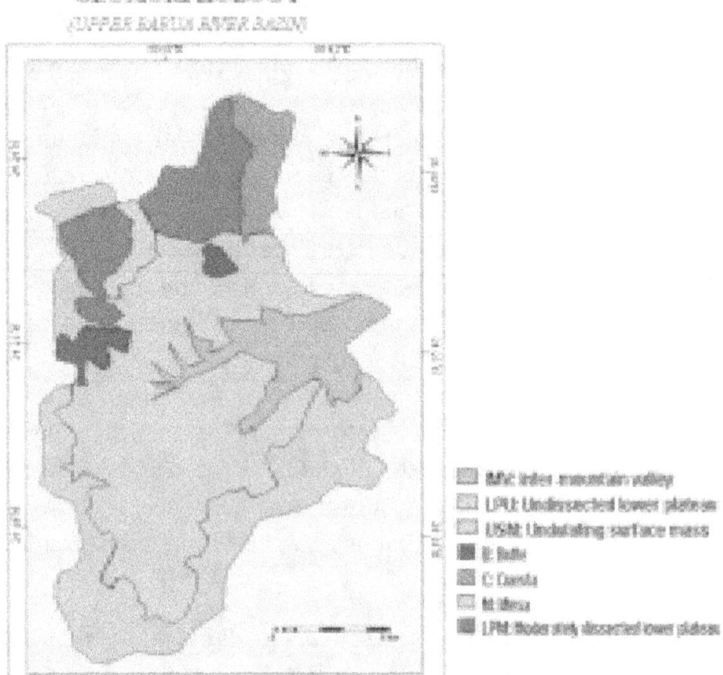

Data Used

1. **Basic Data:** -Survey of India, toposheet 63D/11,scale 1:50000,IRS-1C (LISS-III) FCC of 1998 and 2001, Base map

2. **Ground Data:** -The ground data collection is very much essential to verify the interpreted classes and also to give the training sets to the computer for analysis and also to minimize the fieldwork.

Objective of The Study

The main objective of this study is to learn the methodology of natural resources inventory of a region and to map them on a suitable scale using Satellite Remote Sensing and Geographical Information System techniques, also to identify various natural resources of the area.

Methodology of Natural Resource Mapping

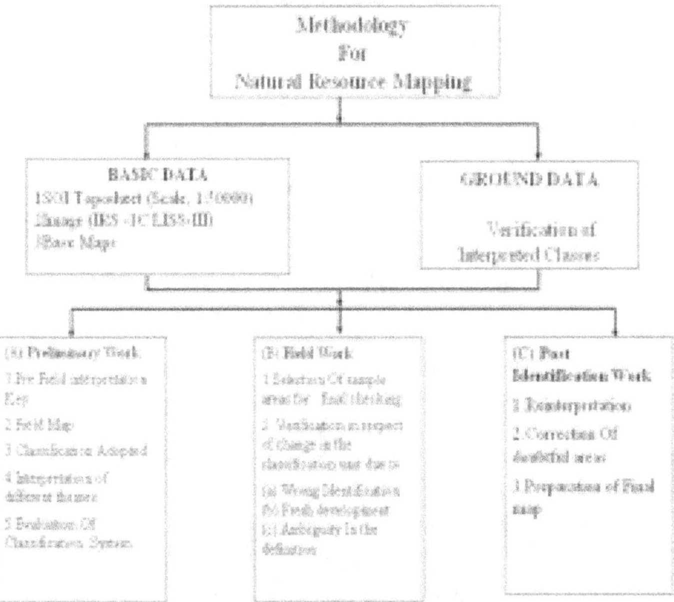

Natural Resource Mapping

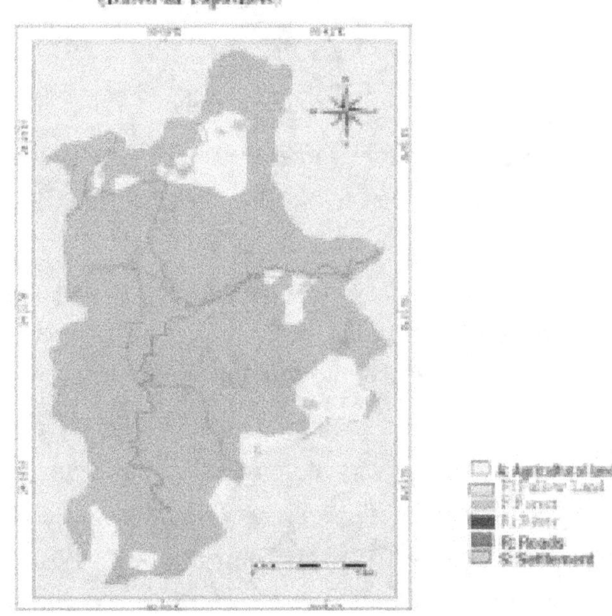

- Area is mainly covered by forest

- Patches of non forest area seen

- Patches contains settlement, agricultural land and fallow land.

- Drained by Barua river in the central of the area.

- Roads connecting the area from all 4 directions.

(b)   Landuse/Land covers (Satellite Image)

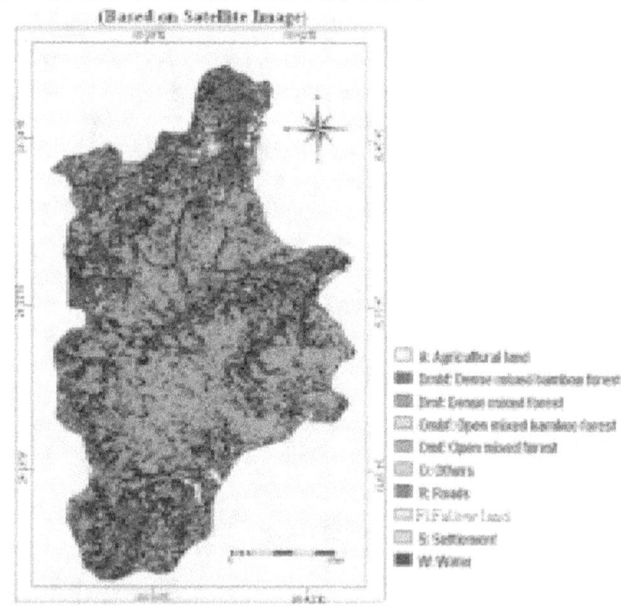

- It also contains forest area mainly

- Patches of non forest area present like settlement, agricultural land & Fallow land

- Scattered patches of water bodies are seen both perennial & non perennial

- Streams & main river drains the area

- Road pixels are seen more comparatively to toposheet

Forest Resource (Toposheet)

- Area is classified into four types of forest

- Non-forest area contains, Open Scrub, Settlement.

- Open jungle covers more area than dense jungle

- Mixed variety of species are found like Teak, Neem, Palash,Babul etc.

Forest Resource (Using LISS-III Image) April 1998

- Four categories of Forest found.

- Open mixed forest largest in quantity
- Area also contains dense forest
- Non-forest area in white patches.
- It contains settlement, agricultural field & fallow land.

Forest Classification (Using LISS-III Image) September 2001

- Most of the area contains Open Mixed Bamboo Forest
- Due to production plantation for commercial use
- Open Mixed Forest is also present in large number.
- Dense forest is less in quantity than Open forest.

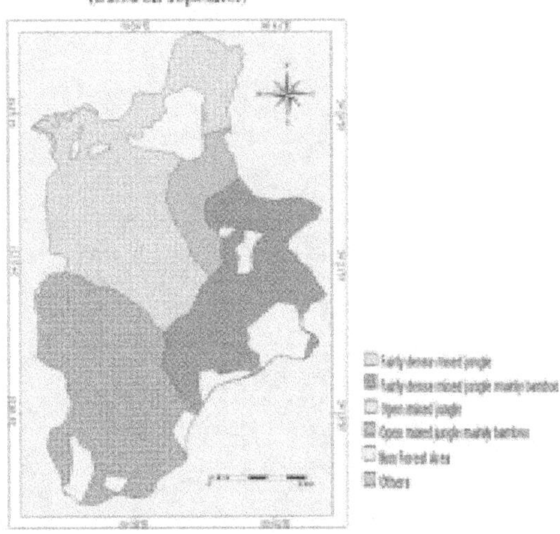

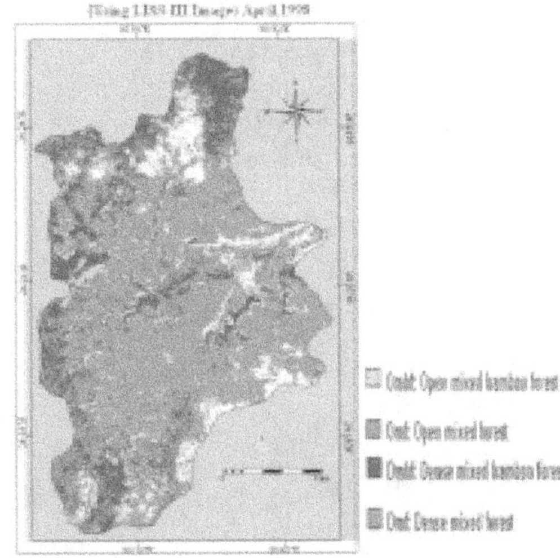

Statistics of Land Cover (Area Sq. Km.)		
Digitally Classified Land cover	Apr-98	Sep-01
Dense Mixed Bamboo Forest	13.93	9.49
Open Mixed Bamboo Forest	2.85	42.99
Open Mixed Forest	60.84	37.4
Dense Mixed Forest	14.3	5.13
Agricultural Land	3.17	2.63
Settlement	2.81	3.23

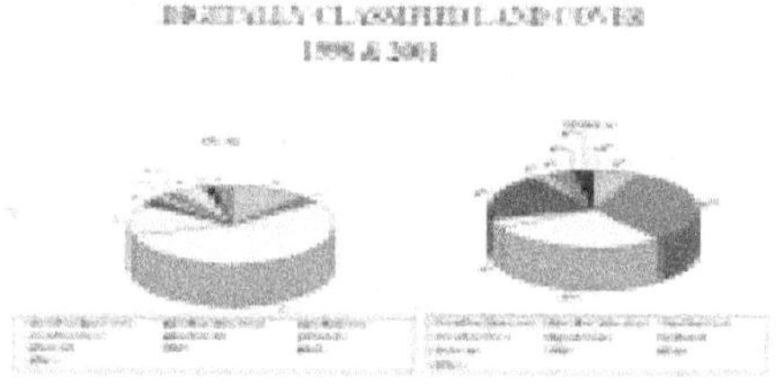

DIGITALLY CLASSIFIED LAND COVER
1998 & 2001

Fallow Land	2.25	4.84
Water	5.12	0.34
Roads	1.88	4.12
Other	3.07	0.07
Total	**110.22**	**110.00**

Statistics Of Forest   (Area-Km²)

Forest Types	Apr-98	Sep-01
Dense Mixed Bamboo Forest	13.93	9.49
Open Mixed Bamboo Forest	2.85	42.99
Open Mixed Forest	60.84	37.4
Dense Mixed Forest	14.3	5.13

FOREST COVER OF THE YEAR 1998 & 2001

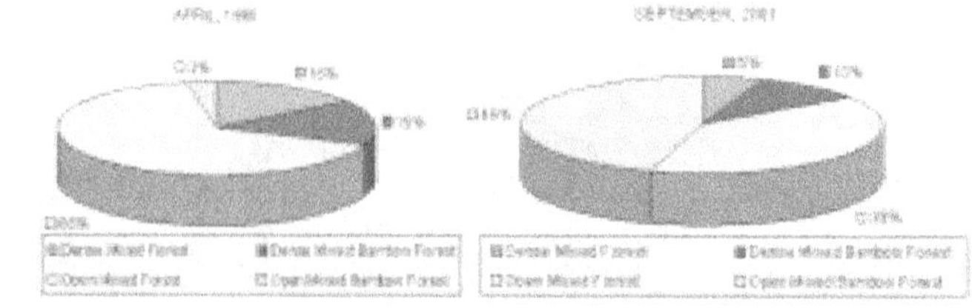

Conclusion

1. Area Mainly Consists Of Forest

2. Degradation Of Forest Due To Deforestation Is Happening In The Area.

3. There Are Some Patches Of Highly Disturbed Areas E. G. Deforestation In Dense Forest.

4. Water Availability Is Moderate.

5. Increase In Agricultural Land And Built Up Areas

6. Production Plantation of Bamboo for Commercial Purpose.

7. Better Accessibility, Due To Development of Roads.

8. Slightly Disturbed Ecology Due To Anthropogenic Activities.

Management practices required

1. Plantation Activities Is Required

2. Mixed Plantation Should Take Place To Escape From The Demerits Of Monoculture.

3. Proper Land Resource Utilization Is Required Through Plantation And Water Storage To Check Soil Erosion And Recharge Of Down Streams Areas

Reference

1. Anji Reddy,M.(2000)Remote Sensing and GIS –An Introduction , J.N.T. University, Hyderabad, India

2. Chawla,A. and Thukral,K.A.(2000)Digital Image Processing of IRS-1B data for Landuse classification of Amritsar,Dept.of Botanical Studies,Guru Nanak Dev University,Amritsar,India.

3. Lillesand,T.M.and Kifer,R.W.(1987)Remote Sensing and Image Interpretation, Second edition,Jhon Wiley and sons,New York.

4. Murthy, K.S.R.and Rao,V.V.(1997).Temporal studies of Landuse/Landcover in Varha River Basin ,A.P.,India, J.Indian Soc.Remote Sensing,vol.25,no.3,pp.145-154.

5. Sudhakar,S.,Kumar,A.,Arrawatia,M.L.and Sengupta,S.K.(1994).Forest cover mapping of east district ,sikkim using IRS-1A LISS II satellite data, J.Indian Soc.Remote Sensing,vol.22,no.3,pp.155-168.

Natural Resource Management For Sustainable Development Using Remote Sensing Technology- A Case Study

V.K. Verma, P.K. Sharma, L.B. Patel, D.C. Loshali and G.S. Toor

Punjab Remote Sensing Centre, PAU Campus, Ludhiana - 141 004

The state of Punjab is intensively cultivated and is contributing large share in the grain basket of the country. However, thirty years after green revolution, we have started experiencing the limitation of intensive resource use without taking care of its long term sustainability. There has been a large scale degradation of land resources due to erosion, salinization and water logging etc. Nearly 25 per cent area of the state is suffering from one or the other land degradation problems. In order to use the land resources judiciously and maintain their productivity, there is a need for sustainable development of these resources. In view of this, the requirement for both accurate and timely information on resources had expanded considerably over the last decade for intedrated resource management with watershed or block as a unit of planning.

An integrated approach using remote sensing offers technologically the appropriate method of studying land and water resources, characterising the coherent agricultural zones and identifying constraints for natural resource management. Integrated studies in selected blocks or watersheds in 175 districts of 25 States of the Country are being undertaken under the Project "Integrated Mission for Sustainable Develompent (IMSD)" coordinated by ISRO, Department of Space. In order to tackle the problems of thick sand cover, soil salinity and water logging and poor quality of ground water in Mansa district of Punjab, India, the integrated resource study for sustainable development was undertaken under the aegis of the project "Natural Resource Development and Management System (NRDMS)" sponsered by the Department of Science and Technology, Govt. of India. The present study reports the resource management needs of Bhikhi block of Mansa district

Study Area

The Bhikhi block of Mansa district (Punjab) covering an area of 402.37 Km2 forms a part of the Indo Gangetic alluvial plain. The Western Himalayas in the north and the Thar desert in the south and south west mainly determine the climatic conditions. The south west monsoon during summer brings the much needed rain bearing depressions from July to September. The area comprises of Indo Gangetic alluvium of Quaternary age. It is an alluvial complex of fluviatile origin deposited by the ancestral tributaries of the Indus River System which include the ancient Satluj River. The area is nearly level, with imperceptible slopes, except for the sand dunes. The study area has the problems of arid climate, thick sand cover (sand dunes), low inherent soil fertility, brackish underground water etc.

Material and Methods
Data Used
i.   Black and white aerial photographs of March, 1988 on 1:50,000 scale.

ii.   IRS 1A/1B LISS II FCC (print form) of May 1992, October 1992, March 1993 and April 1996.

iii.   Survey of India toposheets pertaining to the area on 1:50,000 scale.

iv.   Block map with village boundaries on 1:50,000 scale (published by Director Land Records, Punjab).

Methodology
The IRS 1A LISS-II, satellite data (geocoded false colour composites) of March/April, 1996 generated from bands 2, 3 and 4 were visually interpreted. Simultaneously the black and white aerial photographs on 1:50,000 scale were also interpreted using mirror stereoscope. Various thematic maps viz. geomorphology, soils and landuse were prepared on 1:50,000 scale following the standard procedures outlined in the IMSD Manual. The water samples were collected randomly from the study area and analysed for water quality parameters following standarad procedures to prepare water quality map. All the maps were integrated to come out with a resource constraint map. Based on resource constraints site specific recommendations were made and action plan map generated for the management and conservation of under utilised areas for optimal returns on sustainable basis.

Results and Discussion

Soils

On the basis of physiographic analysis of the satellite data and aerial photographs two major physiographic units viz. alluvial plain and sand dunes were delineated. These were further subdivided based on the tone, texture, pattern, slope and landuse. The alluvial plain is nearly level, intensively cultivated and mostly irrigated. The alluvium was later modified and/or new deposits laid by the occasional shifting of the Satluj river. Originally the Satluj river was an independent river, not belonging to the Indus system, before it joined the Ghaggar in Bikaner. Finally, it abandoned its course in the thirteenth century and joined the Beas river.

Due to change in the river course, the tributaries got silted up in due course and resulted in the formation of sand bars. The sand bar deposits seem to have been modified at a later stage by the aeolian action to form sand dunes locally known as tibbas. On the imagery and aerial photographs these can be seen as elongated stretches along the abandoned or filled up channels or both. The present landscape is the result of combined effect of the localised reworking of the previously existing sand dunes by aeolian activity and mechanical shifting of sand in the recent past. Sand dunes occur as elongated stretches, 2 to 8 metres above the general elevation and their strike is parallel to the prominent wind direction.

The soil samples were collected and analysed for particle size distribution, pH, EC, CaCO3, organic carbon, cation exchange capacity and exchangeable cations. The soils are slightly alkaline (pH 8.5-8.9), having low electrical conductivity (0.05-0.8 dsm-1), low organic carbon (0.01-0.4%) and variable calcium carbonate content. The low organic carbon content (<0.4%) of these soils is due to limited biological activity and rapid decomposition of biomass under the prevalent torric conditions.

Based on the difference in soil texture, drainage and profile development, the soils were grouped into five soil series. The soil - physiographic relationship was established. The soils were classified as per Soil Taxonomy (Soil Survey Staff, 1996) as Ustic Torripsamments (Soil Series 1 and 2), Coarse loamy Ustic Haplocambids (Soil Series 3 and 5), Fine loamy Ustic Haplocambids (Soil Series 4). The final soil map (Fig. 1) was prepared on 1:50,000 scale.

Ground Water Quality

The ground waters of study area are alkaline in reaction (pH >7.0). These waters have varying levels of salinity (0.31 to 2.36 dS m-1), SAR (0.40 to 21.57 [me L-1]½) and RSC (nil to 10.60 me L-1) with a mean value of 1.24 dS m-1, 9.97 (me L-1)½ and —— me L-1 respectively. Sodium is the dominant cation and its value ranged from 0.43 to 34.78 me L-1. Among the anions, HCO3- concentration varied from 1.0 to 13.0 me L-1 with a mean value of 5.88, whereas concentration of Cl- and SO42- varied from 0.75 - 12.25 and 0.21 - 14.17 me L-1, respectively.

Depending upon the EC and RSC values (Sood et al., 1998), the ground waters of the area have been grouped into three ground water quality categories viz. good, marginal (sodic) and poor. The ground water quality map (Fig. 2) of the area shows that the category-I (good) occupy 35 per cent of total geographical area of the block. Since these good quality ground waters have least salinity and sodicity hazard, their use over the years is not likely to be hazardous in the soils having clay content even >30% which are fairly to moderately well drained with water table not shallower than 1.5 metre.

Marginal-sodic ground waters occupy highest area (59%) of the block. These waters are low in EC and calcium but high in sodium and bicarbonate. The indiscriminate use of these waters will result in precipitation of calcium and magnesium as insoluble carbonates thereby, building up higher levels of exchangeable sodium in the soil exchange complex. This will lead to formation of dispersed and relatively impermeable soils. These waters can preferably be used in light textured, well drained and permeable soils, if good quantity of organic manures are added. The poor quality ground waters which are unsuitable for irrigation due to high EC, RSC or both occupy six per cent of total area. These waters should not be used for irrigation purposes otherwise they will cause serious problems of soil salinization and sodification ultimately severely restricting the crop yields and deteriorating the soil healthy.

Landuse

Landuse mapping of the study area has been undertaken using two dates IRS 1B LISS II data. Five major landuse categories at level I were identified and mapped (Fig. 3). These categories have been subdivided based on the differences in tone, texture, pattern, association etc. and eleven categories at level II were identified and mapped. The map was rechecked using April 1996 data. There are 34 inhabited and one uninhabited villages in this block. The study reveals that more than 89 per cent of TGA of the block is double cropped and only 4.32 per cent area is single cropped. The area under settlements and village ponds is 3.46 and 0.36 per cent respectively. Nearly, 2.28 per cent of TGA is under wasteland category which includes salt-affected, waterlogged and sandy areas.

Resource Constraints

The study area has a variety of problems associated with soil and water which are listed below :

    i.   Arid and semi-arid climate

    ii.   Low, erratic, ill-distributed rainfall

    iii.   High wind velocity especially during summer months (April-June)

    iv.   Poor underground water in large area (65%)

    v.   Thick sand cover (sand dunes) in some area.

    vi.   Poor retention of water and nutrients in coarse textured soils.

    vii.   Inadequate and erratic supply of canal water.

    viii.   Poor marketing facilities and absence of processing units especially for horticultural produce.

Based on the information on landuse, soils and ground water quality, the following major constraints in the study area were identified (Table 1) and a resource constraint map of the area was prepared on 1:50,000 scale (Fig. 4)

- Poor ground water quality.

- Coarse textured soils primarily in sand dune areas.

Resource Management

On the basis of resource constraints in the area, action plan for sustainable development has been prepared on 1:50,000 scale (Fig. 5). To address the major problems of the study area, certain measures like sand dune stabilisation, arresting water logging and soil salinity, proper use of poor quality ground water and improvement of soil physical properties etc. are recommended.

A.   Levelling and Stabilisation of Sand Dunes

o   Levelling of low sand dunes and bringing them under agroforestry / agrohorticulture with the provision of drip irrigation system.

o   In situ stabilization of high sand dunes by planting species like Sarkanda (Saccharum munja), Ber (Zizyphus nummularia), Pahari Kikar (Prosopis juliflora), Jand (Prosopis cineraria), Subabul (Leucaena leucocephala), Kikar (Acacia nilotica) along the periphery of fields to check the movement of sand by wind action.

B.   Proper Use of Poor Quality Ground Water : Hazards of irrigation with poor quality waters can be minimised with good soil and water management. The ground water rated as marginal- sodic (RSC 2.5 -7.5 me L-1) can be used safely with recommended gypsum application. However, HCO3- containing water may increase the level of exchangable Na+ in the soil, even if the concentration of Ca2+ and Mg2+ ions exceeds that of HCO3- and CO32- (zero RSC waters), because the precipitation of the cations as insoluble carbonates increases the SAR of the soil solution and hence the level of exchangable Na+. In the regions, where availability of canal water is less, it is some time necessary to use ground water for irrigation with an SAR which is likely to give sufficiently higher concentration of Na+ in the soils, concomitatnly reducing the permeability of the soil to a n unacceptably low level. The exchangable Na+ can still be kept low by adding gypsum (CaSO4.2H2O) either to the irrigation water or to the soil; or sulphur itself can be added to the soil if it contains free CaCO3 which results in the formation of CaSO4. The advantage of adding gypsum is that it helps to maintain the permeability of the surface soil. The ground waters of zone IV, having either EC more than 6 dS m-1 or RSC > 7.5 me L-1 or both, is unsuitable for irrigation. The continuous use of this water can cause secondary problems of soil salinization and sodification which will adversely hamper the crop growth.

Earnest efforts on the part of Agriculture Department are required to educate farmers about the ill effects of continous use of brackish water on the soil environment and productivity.

C.   **Improving Soil Physical Properties :** The coarse textured soils have low water and nutrient retention. Though there is no way to improve soil texture but its moisture and nutrient holding capacity can be increased by the addition of organic manures. Compaction of sandy soils has been found to help greater moisture retention increasing the number of micropores. Use of green manures also helps in improving soil structure.

D.   **Agriculture :** In the study area, the farmers follow traditional cropping rotation of wheat-paddy/cotton irrespective of ground water quality. This practice holds good in areas having good quality ground water only. However, in the areas having saline and sodic ground waters, alternative cropping pattern consisting of suitable crops should be adopted.

- Agriculture I : Sodium tolerant crops like wheat, barley, berseem, cotton, raya, and sugarcane are recommended in areas having light to medium texture soils and sodic ground waters. The amendments like gypsum should be used alongwith the sodic ground water to reduce the harmful effects of sodium.

- Agriculture II : In alluvial plain (double cropped) with medium to heavy texture soils and sodic ground water, high sodium tolerant crops like rice, sugarbeet and bermuda grass are recommended. The ground waters in these areas should be used alongwith gypsum.

E.   **Forestry :**

- Reclamation of salt affected cum waterlogged area and bring them under plantation with species such as Safeda (Eucalyptus spp.), Pahari Kikar (Prosopis juliflora), Kikar (Acacia nilotica) and Neem (Azadirachta indica), which act as biopumps.

- Forestry in low sand dune areas with species like Pahari Kikar (Prosopis juliflora), Kikar (Acacia nilotica), Jand and Ber.

- In alluvial plain areas with brackish ground waters, in addition to the above mentioned forestry species, Eucalyptus and Dek can also be grown, however, if there is assured availability of canal water, agriculture can be the best practice.

F.   Horticulture/Agrohorticulture :

o   Agrohorticulture I: Promotion of horticultural crops like Grapes and Kinnow in marginal lands having good under ground waters, alongwith gram, groundnut and moong.

o   Agrohorticulture II : Promotion of agrohorticulture with fruit plants like Ber, Guava and Amla and oilseeds (Mustard, Toria) in sand dune area having saline or sodic ground waters, after their levelling and/or clearing.

G.   Augmentation of Irrigation Facilities :

o   Assured irrigation with good quality water during the first two years of horticultural plantation.

o   Increase in canal command area with the provision of effective surface and sub surface drainage.

o   Release of additional canal water during the months of April and May and the supply of water should be reduced during the months of June to September. This arrangement in turn will help in reducing the insect and pest attack on cotton crop.

o   Installation of deep tubewells, wherever feasible, by Government agencies to tap good quality deeper aquifer.

o   Application of recommended doses of gypsum should be applied alongwith irrigation waters having moderate to high residual sodium carbonate.

o   Advising farmers to adopt drip irrigation in horticultural crops and sprinkler irrigation in other agricultural lands where good quality underground water is available or have adequate supply of canal water. This will help in efficiency.

o   Alternate furrow irrigation should be advocated in cotton to save water.

o   Power connections on priority for shallow tubewells installed in sweet water zones.

Acknowledgements

Authors are indebted to Department of Science and Technology, Govt. of India for providing financial support to accomplish this study.

*   Sood, A., Verma, V.K., Thomas, A., Sharma, P.K. and Brar, J.S. 1998. Assessment and management of underground water quality in Talwandi Sabo tehsil of Bathinda district (Punjab). J. Ind. Soc. Soil Sci. 46 :421-426

*   Soil Survey Staff 1996. Keys to Soil Taxonomy. Soil Conservation Service, USDA, Washington, D.C.

Table 1 : Resource Constraints in Bhikhi Block

Mapping Units	Cropping Pattern	Geomorphology	Soil Texture	Ground Water Quality	Constraints
1	Barren	Sand Dunes	Sandy	Variable	Coarse Textured Soils, Active Sand Dunes
2	Dominantly Single Cropped	Sand Dunes	Sandy	Good - Low EC, Low RSC	Coarse Textured Soils
3	Dominantly Single Cropped	Sand Dunes	Sandy	Marginal - High EC, Low RSC	Coarse Textured Soils & Saline Ground Water
4	Dominantly Single Cropped	Sand Dunes	Sandy	Marginal - Low EC, High RSC	Coarse Textured Soils & Sodic Ground Water
5	Dominantly Single Cropped	Sand Dunes	Sandy	Poor - High EC, High RSC	Coarse Textured Soils & Brackish Ground Water
6	Dominantly Barren	Old Filled Up Channel	Coarse Loamy	Variable	Soil Salinity Associated With Water Logging
7	Double Cropped	Alluvial Plain	Coarse Loamy To Fine Loamy	Marginal - High EC, Low RSC	Saline Ground Water, Irrigation & Management
8	Double Cropped	Alluvial Plain	Coarse Loamy To Fine Loamy	Marginal - Low EC, High RSC	Sodic Ground Water, Irrigation & Management
9	Double Cropped	Alluvial Plain	Coarse Loamy To Fine Loamy	Poor - High EC, High RSC	Brackish Ground Water, Irrigation & Management
10	Double Cropped	Alluvial Plain	Coarse Loamy To Fine Loamy	Good - Low EC, Low RSC	No Constraint

Natural Resources Data Management System (NRDMS) – A Suite of Technologies for Local Level Planning

**R. Siva Kumar, P.S. Acharya, D. Dutta, M. Prithviraj,
Nisha Mendiratta, Bhoop Singh**
Department of Science & Technology
Government of India, New Delhi

Drawing up strategies for local level (district and below) development that is sustainable, area-specific and take into account the felt needs of the local people is a complex and information intensive task. Major upgradation of the existing data system at the local level is thus essential. Induction of scientific tools and techniques like Geographical Information System (GIS), Remote Sensing and Web Technologies is necessary to make the data system amenable to quick retrieval, flow and holistic analysis. The Government has therefore, encouraged programmes aimed at developing and inducting appropriate scientific and technological tools so as to upgrade the databases and improve data management procedures at the districts. Natural Resources Data Management System (NRDMS), is one such initiative of the Government, conceived and launched by the Department of Science & Technology in 1982 with a distinct focus on organizing local level resource databases and demonstrating their utility in local level planning.

NRDMS Programme

Major objectives of the Programme include:

- Development of district level resource databases on natural resources and other allied sectors based on the concepts of GIS to support local level planning.

- Provide software support for data management, modeling and operation research.

- Promote R&D in spatial data management technologies.

- Training of scientists & potential users.

- Forge linkages with the users at different levels, documentation and dissemination.

Under NRDMS Programme, experimental database centers have been set up in selected districts of the country to develop the local level GIS databases and demonstrate their utility in local level planning. Data sets are collated from different sources like national and state level survey agencies, line departments, and remote sensing (both aerial and satellite) and converted to digital mode for storage on to the databases for integration and analysis. Required R&D back-up of the task of developing databases and software tools, and putting them to use are provided by leading academic/ research institutions and NGOs. Area-specific application studies are carried out to investigate resource related problems at the local level in sectors like watershed management, energy budgeting, infrastructure development, and landslides. Each district center of NRDMS is provided with a minimal set of hardware, software and technical manpower to support the development and maintenance of the databases and data processing. Manpower staffing the Center provide the required technical support to the line department officials in this task.

UNDP has assisted NRDMS in upgrading the NRDMS methodology by way of providing world class state-of-the-art techniques for developing local level GIS databases and the required tools for using those for decision support in different sectors of local level planning.

Various techniques and tools developed under the NRDMS include improved procedures for local level data management. GRAM++GIS package, Decision Support Modules, GIS databases, Resource Profiles and Almanacs, and Tutors on GRAM++ and GIS. The techniques and tools have been tested using data sets from two pilot districts i.e. Bankura in West Bengal and Kolar in Karnataka for demonstration of their use to the Line Department officials. The data flow characterizing the complete system is shown in the attached Figure.

Improved Procedures

Improved procedures of data management involves "Needs Assessment' of the end-users, preparation of a conceptual data model, survey of available data and compilation of metadata, identification of data gaps, and preparation of a detailed design of the GIS database. End-users of the database being the officials from Line Department and Zilla Panchayats at the districts, a detailed 'Needs assessment' study has been carried out to capture their data needs.

Based on the analysis of the outcomes of the Needs Assessment exercise, a list of data items – both spatial and non-spatial – has been identified. Comparing the data requirement with the data available with the survey agencies has helped identify the data gaps in terms of availability (both coverage or temporal) and spatial resolution and formed the basis of suggesting strategies for meeting the gaps. An assessment of the accuracy of the data collected, sources of data, data definition, expected accuracy, timeliness and their updating frequency has been made for compiling the metadata ("data about data") for the databases.

On the outcomes of the needs assessment study, a conceptual design of the envisaged database, called a Conceptual Data Model has been prepared using Entity – Relationship (ER) modeling technique to provide a combined view of the district database. The detailed structure of the database and the associated files were worked out using the Data Model. Data (both spatial and tabular) have been collated or collected from different sources, converted to digital mode and put into the database files to construct the database. A set of about 67 maps and 190 tables constitute the core of a district database. Organising such a huge amount of data on the database and making the database usable to the Line Departments and panchayats require proper data organization and processing tools.

GRAM++ Package

A PC based user friendly, GIS software tool – GRAM (Geo Referenced Area Management) – designed and developed a few years ago to cater to the data management needs at the district NRDMS Centers has been upgraded to GRAM++ with additional features and capabilities on Windows 95/98NT platform. The modules in GRAM++ include Import/Export of different data formats, Map Editing, Vector Analysis, Raster Analysis, Network Analysis, Spatial Query Language support for combined attribute and map based queries, Digital Image Processing, Watershed Analysis, and Map Layouting. The package has interface with MS ACCESS for linking maps with collateral attribute data. GRAM++ has been put to organizing the GIS databases and a series of application studies. A selected set of often – used functions of GRAM++ package has been developed as a collection of programe development libraries that the application developers can use and build GIS application with very small sized code written in Visual Basic, Visual C++ compilers. As per the current standard, these libraries have been created as ActiveX controls and named as GRAM++ tools.

Sectoral Decision Support Modules

Four sectoral decision support software modules have been developed in the identified sectors of water resources management, land use planning, energy management, and infrastructure development. The modules are capable of working on the databases to retrieve the relevant data sets, analyse, and provide information useful for local level planning. While the Land and Water Modules support data processing to generate information on watershed boundaries in a district, watershed-wise surface water availability, crop productivity, biomass yield, and soil erosion status, the Energy Module helps assess the energy demand and supply situations and identify deficit areas in a district/block requiring extra supply of energy. The Infrastructure Module provides tools are locating facilities like health centers, schools, fair price shops and allocating them optimally among villages/ settlements depending on the facilities' capacity. Training kits on GIS and GRAM++ have been developed for training of staff and other end-users to promote the use of the newly developed tools in the task of local level planning.

GIS databases, Resource Profiles and Almanacs

Based on the 'Needs Assessment' study, data sets have been collated from various agencies and Line Departments operating at national, state and district levels. Major national survey agencies that provided data sets for the construction of GIS databases for the pilot districts include Survey of India (SOI), India Meteorological Department (IMD), National Bureau of Soil Survey and Land Use Planning (NBSSLUP), Central Ground Water Board (CGWB), Central Water Commission (CWC), Forest Survey of India (FSI), Geological Survey of India (GSI) and Census of India. Data relating to agriculture, irrigation, cropping patten, road network, health & education facilities, veterinary centers in some instances are collated from the state or district level Line Departments. Limited primary surveys have been carried out in Upper Gandheswari Sub-Basin and Chagalkuta watersheds in Bankura and, Byrasagar and Rampatna watersheds of Kolar district to fill up the data gaps pertaining to the hydrological modeling studies. Similar surveys have been carried out in Kolar for estimating demand and supply of various types of energy. All these data sets are being put together into the database structure prepared out of the Conceptual Data Model based on the "Needs Assessment" study. SOI's topographic map form the cartographic base for the GIS Databases of the districts. Resource profiles and District Almanacs have been prepared using the above data sets for demonstration to the local level officials and use at the district NRDMS centers.

GRAM++ and GIS Tutors

An Interactive Tutor has been developed to demonstrate the use of each and every function of GRAM++ so that new users can get themselves familiar with the functionalities of the Package in no time. A GIS Tutor has also been developed for use in training of staff from the Line Departments and other local level institutions entrusted with the responsibility of preparing local level development strategies.

In addition to the above tools useful in assessing the local natural and social resource endowment and spread of facilities, pilot studies have been undertaken to examine the socio-economic situation of a district vis-à-vis the next higher planning unit – the state. The integrated database for Bankura has been utilized to estimate indicators like migration, literacy (including gender gap), work force (including gender gap), agriculture, industry, employment, access to amenities like drinking water, electricity, and health services which can be compared with the corresponding indicators for the State of West Bengal for drawing up area-specific development strategies. Estimates of the Human Development Index (HDI) of Bankura and the State of West Bengal have been made by combining life expectancy, adult literacy, enrolment ratio and real GDP Per Capital for use in local level planning.

With the above technological tools and resource databases available at the districts, and the institutions of local self-governance (zilla panchayats) in place, it is expected that the process of local level planning will be more scientific and take into account the local resources, and the locally felt needs of the people while drawing up local development strategies. Based on this experience and upgradation of the methodology and tools, NRDMS is now poised towards forging functional linkages with various User Departments and Ministries like Ministries of Rural Development, Agriculture, Water Resources, and Health & Family Welfare. Such linkages are expected to build up the desired spatial data management capabilities into their activities.

Soil Resource Management In Andhra Pradesh

Poonam Malakondaiah, Qamar Iqbal Khan

Department of Agriculture, Govt. of Andhra Pradesh, Hyderabad

Introduction

Soil is the mother earth that provides livelihood for human beings & animals. Morphologically, it is a loose and friable material spread over the land surface, which provides foothold and nourishment to plants. Maintaining soils in a state of high productivity is important for providing people with their basic needs on sustainable basis.

This limited and precious resource is declining in quality and extent due to natural as well as human interventions put forth for food, industrial, social and urban needs. The effects are noticed in physical and chemical characteristics of soil due to erosion, salinisation, nutrient deficiency etc. This is evident from the fact that about 175 Mha (53%) of land area in the country is suffering from various kinds and degrees of degradation (Sehgal, 1990). Andhra Pradesh is no exception to soil hazards, where 55 percent of lands are facing degradation hazard.

The Department of Agriculture has put significant efforts in managing the soils for sustainable productivity. This paper presents glimpses of the soil resources of Andhra Pradesh, their evaluation and the sustainable approaches of the department in maintaining the soil health for improving the agriculture production.

Andhra Pradesh at a Glance

Andhra Pradesh is one of the four littoral states of India, supporting a population of 76.1 million including 55.3 million from rural sector. The state covers an area of 27.7 M ha located between 12° 37′ to 19° 54′ N latitude and 76° 46′ to 84° 46′ E longitude. It is endowed with wide variations in climate, geology, physiography and vegetation, which are reflected in the development of a large variety of soils. There is a transition from tropical to subtropical monsoon climate of semi arid to arid in Telengana and Rayalaseema Regions and humid the sub humid in the Coastal Andhra Region. Average rainfall of the state is 830 mm, varying between 550 to 1250 mm among different regions. The state is divided into seven agro climatic regions based on the precipitation. The average annual evapotranspiration is 1713 mm, with regional variations of 1770 mm in Rayalaseema, followed by 1720 mm in Telengana and 1660 mm in Coastal Region.

Land use in governed by climate, soils and socio-economic conditions. Of the total geographical area, about 40 percent of land is under cultivation. Forests, scrub, cultivable waste, barren and other land use categories cover the remaining. About 40 percent of the cultivated area is irrigated through different sources. A profile of land utilization is shown in Table – 1.

"

Table 1 Land Utilisation in Andhra Pradesh

S.No.	Category	Area (Lakh hectares)	Percentage to Total Geographical Area
1	Total Geographical Area	274.40	100.00
2	Forest	61.99	22.60
3	Barren and Uncultivable Land	20.84	7.59
4	Land Put to Non-Agri. Uses	25.88	9.43
5	Cultivable Waste	7.00	9.55
6	Permanent Pastures and other grazing lands	6.76	2.46
7	Land under Misc. Tree crops, (Groves not included in Net Area Sown)	2.37	1.01
8	Other Fallow Lands	16.79	6.12
9	Current Fallow Lands	35.07	12.78
10	Net Area Sown (including Fish Culture)	97.30	35.46

Source: An outline of Agricultural Situation in A.P (2003 – 04), DES, A.P

Soil Resource Database

The soil resource database generated in the state and the available infrastructure is discussed below.

Soil Survey Information

Soil resource evaluation has been largely neglected not only in Andhra Pradesh but entire India. Hitherto, soil survey was carried out in selected areas like project commands and watersheds. For the first time in the history, country wide soil resource mapping was carried out by the National Bureau of Soil Survey & Land Use Planning (NBSS&LUP) during 1990s at a scale of 1:2,50,000, which gave a generalized information about the soil potential and problems. Although the data does not allow its application at village and farm levels, it serves as an indicator to select the priority sites for further investigation.

The available soil resource data was judiciously utilized for optimal land utilization. The department in co-ordination with the NBSS & LUP came up with generation of district wise land resource atlases containing following 22 themes.

1. Administrative Divisions

2. Soils (Technical Nomenclature)

3. Soils (Traditional Nomenclature)

4. Soil Depth

5. Surface Soil Texture

6. Gravelliness

7. Available Water Capacity

8.   Soil Calcareousness

9.   Soil Slope

10.   Potential Soil Loss

11.   Fertility Status (NPK)

12.   Available Zinc

13.   Soil Degradation

14.   Land Capability

15.   Land Irrigability

16.   Hydrogeomorphology (Groundwater Potential)

17.   Length of Growing Period

18.   Land Use / Land Cover (Existing)

19.   Suitability of 1st Major crop

20.   Suitability of 2nd Major crop

21.   Suitability of 3rd Major crop

22.   Suitability of 4th Major crop.

The two themes, viz, land use and hydrogeomorphology (Groundwater) were generated by NRSA. The land resource atlas is first of its kind in India, which has been circulated to different states by the NBSS & LUP for a similar exercise.

In a move to provide access of the atlas to the field staff, the district wise data is being separated at smaller administrative levels, viz, sub district (division) and mandal (modified taluks) levels, showing all the above 22 themes. Steps are being taken to generate soil maps at a larger scale of 1:50,000 in a phased manner.

Soil Degradation

The major causes for soil degradation are soil erosion and salinisation, besides slight mismanagement by the farmers or other land users. An assessment of soil degradation was attempted (NBSS&LUP, 1996) through soil survey at 1:250,000, which revealed that an area of 15 M ha was affected by various hazards. Water erosion is the major problem causing loss of topsoil and terrain deformation in about 45.5 percent land area. Of this, 20.8 percent is subjected to moderate erosion and 13.7 percent strong erosion followed by slight (8.8%) and extreme (2.2%) erosion classes. This type of degradation is due to high intensity and erratic rainfall, which is further aggravated by the topographic and soil conditions that help in the rapid detachment of soil particles. Added to this is the unscientific soil management and over exploitation due to pressure on land from the ever-increasing population.

Water erosion associated with physical deterioration (crusting and compaction of the surface soil) in the state accounts for about 2.4 per cent area restricted to the Eastern Ghats (North).

Physical and chemical deterioration (crusting and nutrient loss) are prevalent in the laterite areas of northwestern part of Telengana and in Coastal districts, which cover about 1.2 per cent area. Chemical deterioration due to salinity and sodicity covers about 1.8 per cent mostly in the valleys, plains and command areas. Physical deterioration due to water logging and flooding covers about 3.2 percent mostly confined to the delta regions of coastal districts. A very small area about 0.3 per cent is covered under salt flats in the Coastal districts.

With nearly 55 per cent degraded lands in the State there is an urgent need to take up proper soil and water conservation measures, to restore the present degraded forests and pastures, and undertake drainage and other reclamation measures before the soils lose their resilience. A detailed account of soil degradation is presented in Table 2.

Table 2 Soil degradation status in Andhra Pradesh (Area in '000 ha)

Kind of degradation	Slight	Moderate	Strong	Extreme	Total
Water erosion(Wt)	2417.10	5631.48	3772.77	612.37	13423.70
	(8.6%)	(20.8%)	(13.7%)	(2.2%)	(45.5%)
Water erosion (Wt)+ Physical deterioration (Pc)- Crusting	–	670.47	–	–	670.47
Physical deterioration (Pc- Crusting + Chemical deterioration (Cn) nutrient loss	–	–	333.99	–	333.99
			(1.2%)		(1.2%)
Chemical deterioration(Cs) Salinity+ sodicity	142.23	15.80	155.89	203.64	517.17
	(0.5%)	(0.06%)	(0.5%)	(0.7%)	(1.8%)
Chemical deterioration (Cs) – Salinity+ Physical deterioration(Pw)- Water logging	–	–	–	71.48	71.48
				(0.3%)	(0.3%)
Physical deterioration(Pw) – water-logging+flooding	891.80	–	–	–	891.80
	(3.2%)				(3.2%)
Salt flats(Z)	–	–	–	80.45	80.45
				(0.3%)	(0.3%)
Total area	3451.10	6317.53	4262.45	960.98	14992.64
	(12.5%)	(20.3%)	(15.4%)	(3.5%)	(54.7%)
Miscellaneous land type (Rock Land)					2798.39
					(10.1%)

Source: NBSS&LUP (1996).

Soil Testing

The soil resource evaluation discussed above, helps in assessing the inherent capability of soil to support different types of uses and conservation practices. Having decided a specific land use or cropping pattern, it becomes necessary to evaluate the soil fertility status in meeting the crop nutritional requirement. It may be remembered that a highly productive soil would not yield to its capability unless the nutritional corrections are attempted. Soil fertility is a dynamic characteristic unlike other soil characteristics like soil texture, structure etc. As such variation in soil fertility is observed both spatially (farm to farm) and temporally (time to time) due to difference in the soil reserves crop specific demands and the farmers ability and capability to apply soil nutrient. It is therefore necessary to conduct soil test in the laboratory periodically at close intervals of space and time.

Soil Test Infrastructure

Owning to the importance of soil test data for efficient and economical fertilization, the department has set up a network of 82 soil testing laboratories with an annual capacity of analyzing 5 lakh soil samples. The type of laboratories and their capacities are furnished in Table – 3.

Table 3 Soil Testing Infrastructure in Andhra Pradesh

S.No	Type of Laboratory	Nos	Annual capacity of each lab (samples)	Total Annual Capacity (samples)	Facility
1	Regional & District level labs	23	14500	333500 Macro-nutrients 44,000 Micro-nutrients	Macro & Micro nutrient &water analysis
2	Mobile Labs	4	8625	34500	Macro nutrients
3	Agrl. Market Committee level labs	55	2400	132000	Macro – nutrients
	Total	82	-	500000	

Soil Test Data

In a true sense of the 10 million farm holdings in the state, every farm holding needs to be covered by soil testing periodically for optimal fertilizer application. At this rate the annual capacity of soil testing has to be enhanced from 0.5 million to 2 million so as to cover every farm holding in a period of 5 years. Earlier soil samples were collected from selected farmer fields, due to which complete fertility status of the state could not be assessed, leaving gaps in several areas. It was therefore planned to have a systematic coverage of soil test data of entire cropped area of about 10 million ha in the state by collecting one million soil samples in a period of two years (2002-03 to 2003 – 04) from a homogeneous block of 10 ha. Accordingly, 1.10 million soil samples were collected, by reducing the size of the block from 10 ha to 5 ha in problem soil patches.

The samples were analyzed for EC, pH, OC , N,P and K. In addition to the macronutrients, selected samples were analyzed for micronutrients viz, Zinc, Iron, Copper and Manganese. The analytical data revealed that soils were in general deficient in nitrogen and phosphorous and high in potassium content. Similarly among the micronutrients, zinc was deficient in most of the soils, whereas, Iron, Manganese and copper were found to be sufficient in most of the soil samples. The soil test data in respect of macronutrients and micronutrients is presented in Table 4 & 5 respectively.

Table 4 Soil Test Data for Macronutrients during 2003-04 and 2004-05

No. of samples analyzed	Nitrogen			Phosphorous			Potassium		
	Low	Medium	High	Low	Medium	High	Low	Medium	High
1100000	769845	203964	126187	514749	390969	154282	126187	301105	572708
	70%	19%	11%	48%	36%	15%	10%	28%	52%

Table – 5 Soil Test Data for Micronutrients during 2003-04 and 2004-05

S.No.	Micro-nutrient	No. of samples analyzed	Below Critical limit	% Percent	Above critical limit	Percent
1	Zinc	30744	17832	58	12912	42
2	Iron	30744	4620	15	26124	85
3	Manganese	30744	7744	25	23000	75
4	Copper	30744	2544	8	28200	92

Distribution of Soil Health Cards

The soil test data of the 1.1 million soil samples was communicated to the farmers through soil health cards, prescribing the suitable crop and optimal fertilizer doze including organic matter (FYM) for the crop preferred by farmer.

Soil Health Management

The Department has several programmes to maintain the soil health for sustaining the productivity. Important measures undertaken are as follows :

Reclamation of Alkali Soils

Soil test data indicates the development of alkalinity. The soils affected with this hazard are reclaimed by treatment with Gypsum, which is supplied at 75% subsidy.

Correcting Zinc Deficiency

Deficiency of Zinc (Micronutrient) has been noticed in recent years due to heavy exploitation by the high yielding crops. The department is providing ZnSO4 fertiliser in Zinc deficient soils at 50% subsidy.

Correcting Sulphur deficiency in Groundnut crop

Sulphur is generally deficient in groundnut growing areas due to heavy exploitation by the crop. Gypsum is being supplied to growing farmers at 75% subsidy.

Vermicomposting

Organic matter deficiency is a common constraint in agriculture mostly in the semi -arid region. Efforts are being made to improve the organic matter content in soil through supply of vermicompost units comprising of shed and worms at 75% subsidy. .

Green Manuring

A green manure crop grown prior to the main crop and incorporated in the field would enhance the organic matter content in soil. Pillipsera, Dhaincha and Sunhemp seeds are being supplied to the farmers at 50% subsidy.

Impact Of Interventions On Soil Health

Significant impact was noticed by way of different interventions carried out for soil health management. A few examples are furnished below:

Soil Test based Fertilizer Application

Soil Test data was found to be effective in fertilizer application, which resulted in optimum yield with maximum benefit. A case study of fertilizer application in groundnut crop is presented in Table 6.

Table. 6 Soil Test based Fertilizer Application in Groundnut crop

Name of the farmer   -   **Narayan Reddy**

Village   -   Bukkarayasamudram

Mandal   -   Bukkarayasamudram

District   -   Ananthapur

S.No.	Treatment	Fertiliser doze (Kg/ha)			Pod Yield (Q/ha)
		N	P	K	
1	Farmers Practice	50	40	26	22.5
2	Blanket recommendation	30	40	40	24.5
3	Soil Test recommendation	31	14	56	46.5

Land Reclamation :

Gypsum was applied as an amendment in alkali soil, which resulted in increased yield of paddy crop. A case study is presented in Table 7.

Table 7   Benefits of Gypsum Application for reclamation of soil

Name of the farmer	-	D.Janardhan Reddy
Village	-	Cherla Bhutkur
Mandal	-	Karimnagar
District	-	Karimnagar
Crop	-	Paddy
Gypsum applied	-	5 tons / ha
Yield before treatmen	-	22 Q / ha
Yield after treatment	-	35 Q / ha

Application of Gypsum for Sulphur

Gypsum was applied in groundnut crop @ 5 quintals /ha for correction of sulphur deficiency, which enhanced the yield by 20%. The data of a specific field is given in Table 8.

Table 8 Application of Gypsum for Sulphur Nutrient

Name of the farmer	-	M. Venkateshwara Rao
Village	-	Koppaka
Mandal	-	Eluru
District	-	West Godavari
Crop	-	Groundnut
Gypsum applied	-	5 Q/ ha.
Yield before treatment	-	18 Q / ha.
Yield after for treatment	-	21.6 Q / ha.

Vermicomposting

Vermicompost developed and applied in paddy field resulted in the increase of yield by 25%. The data is presented in Table 9.

Table 9 Effect of Vermicomposting

Name of the farmer	-	M.Sita Ram Reddy
Village	-	Baramgudem
Mandal	-	Tadepalligudem
District	-	West Godavari
Crop	-	Paddy
Yield before treatment	-	52 Q / ha
Yield after for treatment	-	66 Q / ha

Green Manuring

Green Manure seed of Dhaincha was sown prior to paddy crop and incorporated in the field that gave a yield hike of 4.5 Q / ha. The supporting data is presented in Table 10.

Table 10 Effect of Green manuring on Paddy Crop Yield

Name of the farmer	-	J. Mallaiah
Village	-	Chamanpalli
Mandal	-	Karimnagar
District	-	Karimnagar
Dhaincha seed applied	-	30 Kg/ha
Yield before green manuring	-	19 Q / ha
Yield after green manuring	-	23.5 Q / ha

Future Measures

There is a tremendous scope of sustaining the soil health and optimizing the productivity, by adopting the following measures, which are suggested herein based on the past experience.

- Creation of soil resource database at 1:50,000 scale for all the districts in a first phase, followed by 1:10,000 or 1:12,500scale.
- Analysis and interpretation of soil resource data in GIS for various crops.
- Updating of soil resources data periodically.
- Use of satellite data for monitoring the soil resources.
- Generation of farm wise soil fertility data and digitization on village maps.
- Strengthening of soil testing labs by increasing the number of labs and creating facility of sulphur and micronutrient analysis in all the labs.

- Up-linking the soil resource and fertility data in a departmental website, enabling access to all the farmers / functionaries.
- Use of sophisticated and quick analytical equipment.
- Imparting training in soil testing, remote sensing and GIS to the staff.
- Supply of Portable Soil Testing Kits to each village for instant assessment of soil fertility.

Conclusion

Soil is a non-renewable and non-elastic resource, which needs utmost care to conserve and restore it for the future generations. Systematic approach to soil heath management, would, not only meet the food requirement of the growing dependents, but also influence in environmental sustainability.

References

Sehgal, J.L (1990) Soil Resource Mapping of different States of India – Why & How? NBSS Publ. Soil Bull.23

NBSS & LUP (1996) National Bureau of Soil Survey & Land Use Planning Soils of A.P, NBSS, Publ. 69 b.

Ground Water Assessment Condition Using the Satellite Data

***Abhishek Srivastava, **Shweta Srivastava, *Abhishek Misra**

*Remote Sensing & GIS Division (Student University of Allahabad), **Swarnpath GIS Forum

ABSTRACT

Water is a precious and most commonly used resource. Surface water resources being exploited from time to time may become short of supply or may not be easily available at site. Fresh water being one of the basic necessities for sustaneous of life, the human race through the ages has striven to locate and develop it. Water is indispensable for life and more so far man. The need for water is felt more and more for better living with modern services. The demand for water for irrigational and industrial complexes also increased correspondingly to meet the requirement of the growing population. India is bestowed with great rivers, the unutilized flows of which are around 1677 billion cubic meters. Groundwater resources availability depends on the recharge & the with drawl from in area. The main sources of recharge are the rainfall, the seepage from canal network & the potential recharge from shallow water table area. An attempt has been made to estimate the groundwater resources potential & balance for future development of the city area. By adopting a GIS platform the result obtained will be faster and more accurate. Till recently, ground water assessment was based on laboratory investigation, but the advent of Satellite Technology and GIS has made it very easy to integrate various databases.

Water is the most valuable & vital resources for sustenance of life and also for any developmental activity. Like many city of the country especially the cities of M.P. the Serainji area has also limited surface water supply & hence it faces an acute water problem. The area represent land of gentle slope responsible for infiltration and ground water recharge. Present study suggests the need for an integrated approach for assessments of ground water resource and surface water supply in city using Remote Sensing data & GIS techniques. This paper also presents results of the study of ground water assessment & water supply by integrating &analyzing various thematic maps derived from both remote sensing & conventional methods using GIS. GIS is not a new concept or technique, it was introduce in Europe & America in the early seventies much before the availability of GIS software. Development activities as whole irrespective of their scale & magnitude effect the environment in short term as well as long term. Environmental problems have attracted the attention of planner's policy maker & politician. It is almost a matter of concern if each and every one to think about the increasing water pollution.

Hydrogeomorphological and lineament maps have been prepared using IRS 1B LISS-II data by visual interpretation. Topographic information has been collected from SOI toposheet at 1:50000 scale & TIN has been generated from elevation contour at 20m interval and spot elevation. A slope map has been prepared from TIN. Surface drainage map has also been prepared from SOI toposheet and satellite data on 1:50,000 scales.

Introduction

Groundwater is a most important natural resource required for drinking, irrigation and industrialization. The resource can be optimally used and sustained only when quantity and quality of groundwater is assessed. It has been observed that lack of standardization of methodology in estimating the groundwater and improper tools for handling the same, leads to miscalculation of estimation of groundwater. It is essential to maintain a proper balance between the groundwater quantity and its exploitation. Otherwise it leads to large-scale decline of groundwater levels, which ultimately cause a serious problem for sustainable agricultural production. A possible solution for such problems is micro level planning, and use of standard methodology for assessing the groundwater. In recent years micro level planning has gained acceptance, since it can be locally applied and readily managed by self-sufficient rural governance.GIS enables effective and efficient manipulation of spatial and non-spatial data for scientific management of watershed and develop alternatives of development model for the benefit of local people .IRS data has been found to be extremely useful in identifying linear features such as faults and fractures, which are ideal localization points of groundwater especially in hard rock terrains.One of the grates advantages of using Remote Sensing data for hydrological modeling and monitoring is its ability to generate information in spatial and temporal domain which is very crucial for successful model analysis ,predication and validation .The GIS technology provides suitable alternate for efficient management of large complex database.

The application of GIS technology in water resource investigation requires the design and development of a methodology to analysis water resource and its components with its complex ecological and socio-economic inter-relationship .It is necessary to translate system dynamics into predictive statement for different spatial and temporal scales. Remote Sensing & GIS are playing a rapidly increasing role in the field of hydrology and water resources development. Remote sensing provides multi-spectral, multi-temporal and multi-sensors data of the earth's surface. One of the greatest advantages of using remote sensing data for hydrological investigation and monitoring is its ability to generate information in spatial and temporal domain, which is crucial for successful analysis, prediction and validation.

Study Area

The study covers the Vindhyan super-group in district Satna of M.P. The area is located by 80°48'00" to 80°59'00"E longitudes and 24°11'00" to24°18'00"N latitudes. An altitude varies from 329 to 648 m. The total area is 129 sq.km. Topographically the area has moderately undulating features .It is drained by various order of drainage. Tone is the main rivers of these area .The different types of soils in the area are sandy and alluvial.

Geology And Geomorphology

Geomorphological study of the area was carried out through remote sensing technique to find out the occurrence and destroy button of different geomorphic units .The understanding of Geomorphology is very important for the present study because topographic feature or relief features of the earth surface with composition is taken into account for understanding the hydrological condition .The broad objective dealt in geomorphology is dimension and scale of relief feature or landforms process acting on the landforms. Geologically the area comprises of upper Vindhyan formations consisting of sandstone, quartzite and shale (CGWB, 1985). Vindhyan formation is overlain by quaternary alluvium, which was deposited on the eroded basement. Upper Vindhyan formation represented by kaimur series are divided into two groups, the upper & lower.

Methodology

The interpretation work was carried out systematically in the following order:

- Satellite interpretation
- Ground truthing for mapping and other details
- Base map preparation through topographic maps.
- Incorporation of other thematic details through topographical maps for final map preparation.

Data Used

- Satellite Data IRS 1C LISS III FCC
- Ancillary Data: - Survey of India toposheets No.63D-15/63 D-16.
- Base Map
- Existing thematic geological map
- Field data.

Objective

- To evaluate and understand the surface and subsurface condition, a detailed study of Serainji Watershed.
- To map the various landforms and associated features in the study area.
- To integrate the thematic data using GIS to evaluate the groundwater resource of the study area.
- To asses the impact of population explosion on drinking water scenario of Study area.
- Special variation of water related problems in Study area and how these can be solved.

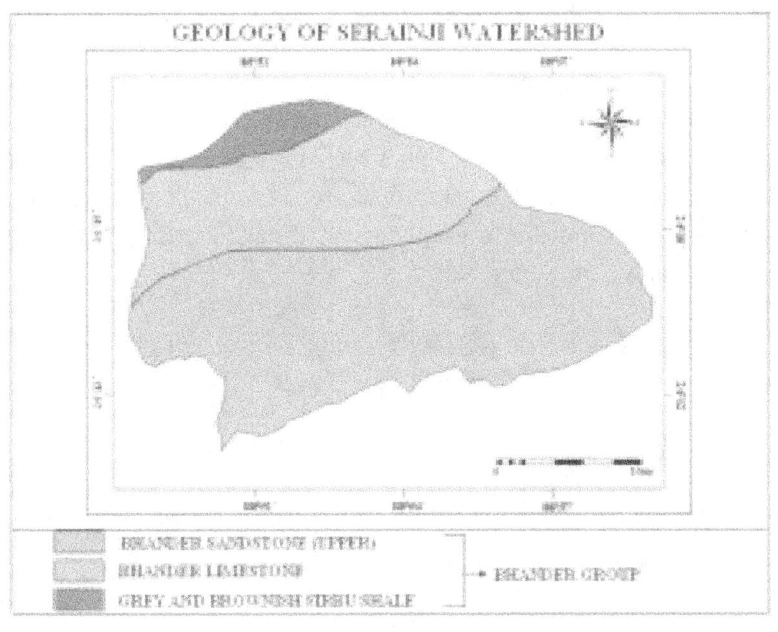

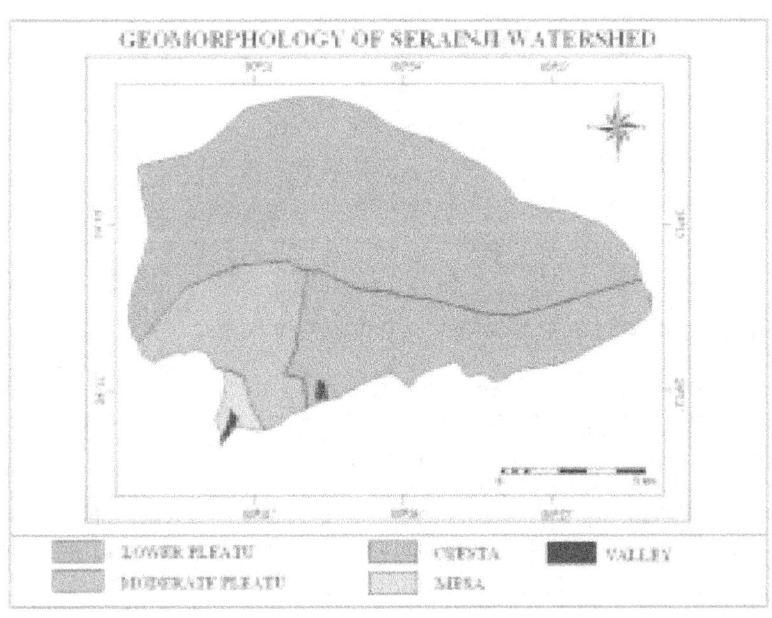

Fig. 6.1 Geological & Geomorphology Map of the Area

Analysis

Thematic map such as geology, slope, geomorphology, landuse and lineament buffer are generated after the visual interpretation were closely scrutinized and then digitized .The digitized maps were then edited and rasterised to suite has an input variable in GIS analysis. This whole process has given an output of digital database required of the study. The whole study area is divided into 5 category i.e. Excellent, VeryGood, Good, Moderate, Poor. The occurrence and movement of groundwater in an area is controlled by various factors it in the area. Therefore each parameter is assigned a weightage depending on its influence on the movement and storage of groundwater.

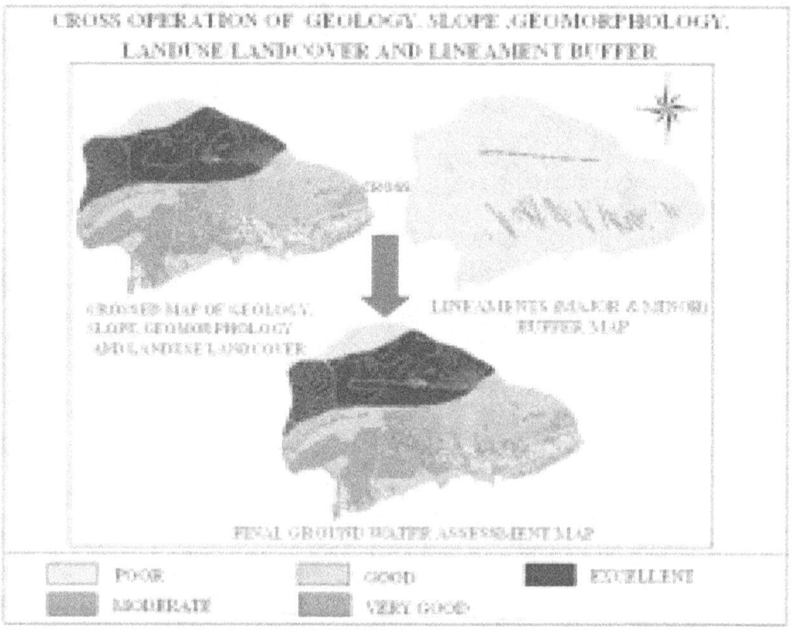

Fig. 6.2 Equal Weitage Map of Geology, Landuae/Landcover, Slope, Geomorphology

Groundwaters Recharge Practices

- Groundwater Recharge scheme should be taken up in city areas to alleviate the deteriorating and declining groundwater condition.

- To recharge the groundwater small pits having bed of permeable material may be made mandatory in the open space/lawn of each dwelling unit so as to face the rooftop water percolate into the groundwater.

Conclusion and Discussion

The category is purely comparative pertaining of the study area. No quantitative assessment has been done due to non –availability of yield data .The excellent groundwater potential area lies geologically in Bhander limestone and around 10m buffer of the lineament in the bhanders and stone area, slope is flat and gentle, geomorphological landform are lower plateau .All these factors are contributing to form good aquifer zone.

Reference

- CGWB, 1985, Report on hydrogeology and Groundwater potential of Mirzapur district.

- Issue in water resource development management and the role of Remote Sensing technical report ISRO, NNRMS-TR-67-86by P.Balkrishna.

- Lecture and case studies of training course on application of Remote Sensing and GIS in groundwater exploration 1993.

- K.S.R Murthy 2000 ground water potential in a semi arid region of. A GIS approach international journal of Remote Sensing Vol. 21no. 19 1867-1884.

- Issue in water Resource development management and the   of Remote Sensing technical report ISRO - NMRMS - TR - 67-86 1986 by Balkrishna

- Current science, special section: IRS-1C Volume -70 No. 7, 10 April 1996.

- www.gisdevelopment.com

Sustainable Management of Wetlands
A Case Study of District Kota Rajasthan

Babu Lal Sharma, Pooja Puar

Department of Geography, P.G. Govt. College, Kota

In todays developing world wetland ecosystems are among the most threatened of all environmental resources. The maintenance of a sustainable flow of benefits derived from the wetland stock is the key issue for maintaining balance between socio economic demand and ecological balance.

"Wetlands[1] are the lands transitional between terrestrial and aquatic system where the water table is usually at or near the surface or land is covered by shallow water." Wetlands in India[2] are distributed in different geographical regions, ranging from the cold arid zone of Laddakh to wet Imphal, from the warm and arid zone of Gujarat – Rajasthan to the tropical monsoon based regions of Central India and the wet and humid zone of Southern Peninsula.

Wetlands are often rare and even unique ecosystems. Valuation[3] of wetlands is an anthropocentric activity, i.e. it means different things to different people depending on their need and greed, and cultural, economic and educational backgrounds. It can be categorized in four divisions:

Global Values

Global Values includes those of widespread significance such as the contribution of wetlands to the mosaic of ecosystem, which maintains global diversity, and the special value of some wetlands as ecotones between dry land and open water.

Functional Values

Functional values includes the ability of wetlands to ameliorate the forces of flood waters and their use in flood control management; water supply and ground water replenishment and the effects of wetlands on micro climates.

Habitat Values

Habitat Values provides habitats to some of the rarest animals and plants and a conducive environment for a wide variety of other organisms.

Anthropogenic Value

Anthropogenic Value includes extrinsic and intrinsic values.

Extrinsic : that cater mainly to government and private organizations that exploit wetlands for major commercial purposes.

Intrinsic : are of direct value to the people who live near the wetlands.

Wetlands have a more complex and fragile ecosystem, they do not have a 'self cleaning' ability, and therefore they readily accumulate pollution; as a result various ecological problems occur in wetlands like eutrophication, toxic contamination, accelerated sedimentation, excessive water diversion, fish depletion, introduction of exotic species and habitat alteration.

Objective and Methodology

Due to lack of awareness wetlands of District Kota has come under the 'Low Protection – High Threat' category, the main objective of the present paper is to focus on some of the conservation efforts and its implementation with sustainable management strategies. To prepare the present paper timely field study of wetlands of District Kota was conducted and through the collection of primary and secondary data and a questionnaire, we thoroughly observed and studied the threats faced by the wetlands.

Hypothesis

* The number and area of natural wetlands to a greater extent have been reduced because of the encroachment in these areas.

* The ecological importance of the wetlands have fallen victim to shrink as a result of population explosion, merging of the cultivated areas adjacent to them.

* The aquatic and marshland flora have been reduced because of the deposition of solid wastes, land, soil and water pollution in these areas.

Study Area

District Kota, is selected as the study area for the research work, it forms the southeastern part of Rajasthan State; lies between 24°25' and 25°51' North Latitude and 75°31' and 77°26' East Longitude. The total area of the district is 5767.97 sq. kms, which is 1.68 percent of the total area of the State. The rocks of Vindhyan System cover the major part of the district.

The soils of Kota region are clay loam to clay containing 30-40 percent of clay in surface soil. District falls in semi arid tract with medium rainfall, the average annual rainfall is 88.56 cms, is received in the monsoon season (July-Sept), distribution of the rainfall is highly erratic and uneven. Mean annual maximum and minimum temperature of the district is 40°c and 29°c respectively.

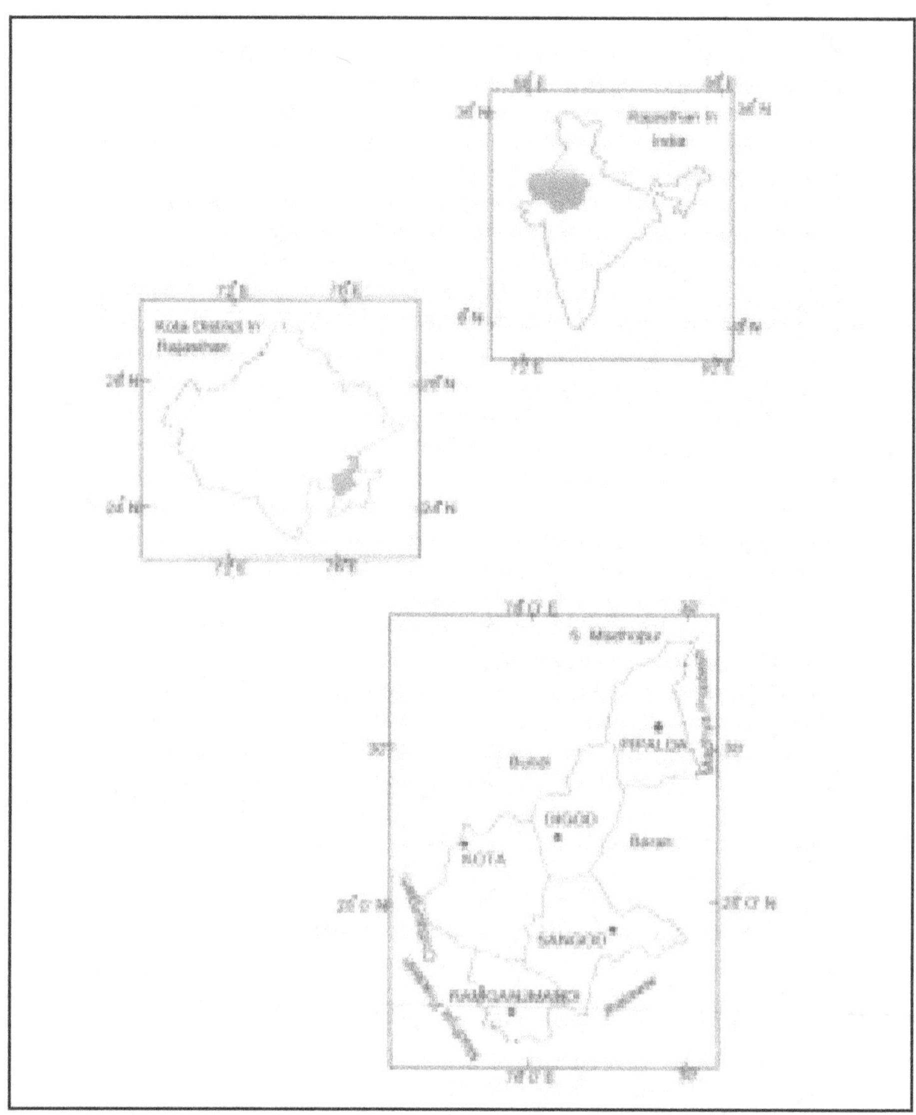

Fig. 2.1 Key Map of District of Kota

River Chambal originating from the hills of Western Madhya Pradesh is the only perennial river of the district. The district attracts the Jawahar Sagar Dam, Kota Barrage and Darrah Sanctuary. Kota city is known as the Industrial city now famous as an Educational center of India.

Status of Wetlands in District Kota

Wetlands of district Kota are precipitation dominated wetlands, it is bestowed with twenty-six natural wetlands and five man made wetlands.

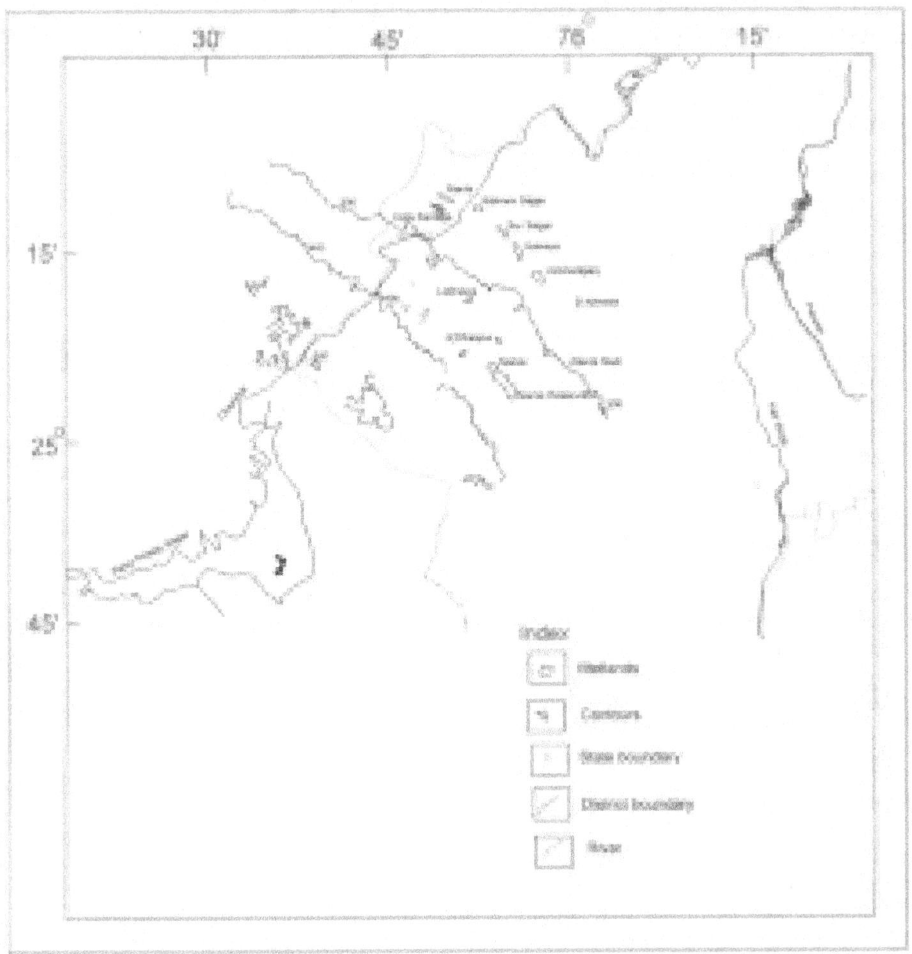

Fig. 2.2 Key Map of District of Kota

Natural Wetlands[4] : Kala Talab, Soor Sagar, Raipura, Kishanpura, Dungarja, Seemliya, Digod, Udpuria, Rajpura, Bandahera, Relavan, Chaarchumma, Boranbaans, Girdharpura, Mandirgarh, Rauntha, Haripura, Mandana, Kaliya Kheri, Polai diara, Brijpura, Kishore Sagar (Darrah) and Sawan Bhadho.

Man Made Wetlands[4] : Alania, Ranpur, Lakhawa, Abheda and Kota Barrage.

During the field study it was observed that there is a high rate of destruction and degradation of wetlands of district Kota. Factors responsible for its loss are: Population Explosion, High Consumption Rates and Intensive Use of Technology for conversion to agriculture use.

We can say that the major threats of Kota wetland sites in order of frequency are from: Fishing (90%); Irrigation (62%); Grazing Animals (56%); Plant Collection (60%) and Domestic Use (49%).

There is no single factor that is associated with the loss of wetlands rather; there are a multitude of factors, mostly anthropogenic, responsible for the loss of wetlands:

Reasons :

(a)   Lack of Awareness
(b)   No Policy
(c)   No Wetland Act
(d)   No Institution
(e)   Over Population

Encroachments :

(a)   Drainage
(b)   Devegetation
(c)   Filling in
(d)   Conversion
(e)   Construction Activities
(f)   Ground Water Extraction
(g)   Discharge
(h)   Digging of Wetlands
(i)   Animal Grazing

▼

Effects :

(a)   Subsidence / Sedimentation
(b)   Vegetation Succession
(c)   Pollution
(d)   Eutrophication

▼

Habitat Loss :

(a)   Destruction
(b)   Deterioration
(c)   Depletion
(d)   Diminution
(e)   Degradation
(f)   Disappearance

Fig.2.3 Dumping of Solid and House hold Wastes

Fig. 2.4 Dumping of Coal Spur and Discharge of Chemicals

Fig. 2.5 Pollution of Wetland water through domestic and cattle use

Efforts Towards Sustainable Management of Wetland Resources

Wise use[5] of wetlands, as stated by the **Ramsar Convention,** involves conservation of the ecosystems while ensuring benefits to the local communities, particularly to the weaker sections of the society on a long-term basis.

Partridge was right, when he said, **"It is a fundamental paradox of our age that scientific knowledge and discipline, supplemented by critical moral sense and passionate moral purpose, will be needed to save the future."**

Experience shows that the values of wetlands can be reduced considerably when they are managed without adequate knowledge and long term strategy. Some of the concepts put forward for the management of wetlands is:

(a)   Ecosystem Management
(b)   Sustainable Management, also known as 'wise use' concept
(c)   Adaptive Management and
(d)   Participatory Management

Ecosystem Management

Ecosystem Management is the scientific way of maintaining wetlands. Like all other organizations in the district, it is very essential for the wetland management agencies to address their need effectively then only their needs will be identified clearly and priorities will be assigned according to the importance.

As surveyed and predicted that the wetland resources of the district have become scarcer, it is very essential to apply more stringent top - down, command and control type of management options.

The golden rule is that 'natural resource management should strive to retain critical types and ranges of natural variation in ecosystems'. That is, management should facilitate existing process and variability's rather than changing or controlling them. By doing so, ecosystem resilience and the organizing processes and structures of ecosystem will be maintained thus better serving, not only the natural functions and species diversity of those systems but also the long term (although not necessarily short term) interests of humanity.

Adaptive Management

Adaptive Management approach views management as an experiment – a process of learning from experience where by we increase our understanding of the reciprocal relationship between natural systems and social systems across time and space.

Sustainable Management

Sustainable development of wetlands encompasses both the concepts of ecosystems and adaptive management. Sustainable Wetland Management[6] has received most thought which finally led to the adaptation of the principle of the 'wise use' of wetlands. This principle states that the 'wise use of wetlands is their sustainable utilization for the benefit of mankind in a way compatible with the maintenance of the natural properties of the ecosystem.' Sustainable utilization is defined as 'human use of a wetland so that it may yield the greatest continuous benefits to present generations while maintaining its potential to meet the needs and aspirations of future generations.'

Hence, for the sustainable management in the district the 'wetland management agencies' should follow the following guiding principles:

* Maintaining the essential values of wetlands

* Preserving the multi functionality of wetlands

* Taking account of the inter relationship between wetlands and other ecosystems, such as:

    - Socio Economic Link
    - Hydrological Link
    - Ecological Link

* Involving rural, wetland dependent communities

* Integrating conservation and development.

Monitoring of the Wetlands

Emphasis should be laid on the survey and mapping of the wetlands by using Remote Sensing technology in the district; these surveys carried out for inventory will not only provide information on biodiversity and ecological characteristics of the wetlands but also provide the information about the different wetland sites and the changes that have taken place over a period of time with regard to their area, vegetation cover, faunal distribution, siltation, encroachment and over all drainage pattern.

Survey and Mapping

A physical survey and mapping of each wetland and delimitation of boundaries is an urgent necessity. During the field study it was found that natural and man-made wetlands are encroached upon and have been transformed into the dumping sites and other recreational activities. In this respect satellite mapping of the area is not only a benchmark but also an aid to conservation.

People's Involvement

A good plan to protect wetlands is not merely through legislation, but mainly through the involvement of locals through awareness development and voluntary involvement is needed.

Conclusion

Long before **Mahatma Gandhi** said, **"Earth can provide for all the needs of humans, but not for their greed's"**. But today the increasing materialistic desire, uncontrolled economic activities has directly undermine the potential for development by over exploiting wetland resources and drastically reducing the carrying capacity of the wetland ecosystems.

Kautilya, the wise minister of Chandragupta Maurya said, **"Stability of an empire depended on the stability of its environment"**. In other words, he emphasized the concept of sustainable development, now it's the time to realize and ponder on the saying of Kautilya.

Reference

1. Cowardin et.al. (1979) : A Report on The Wetland Conservation, SAC – 1991.

2. Garg, J.K., T.S. Singh and T.V.R. Murthy, (1998), 'Wetland of India', Nationwide wetland mapping project report, Ahmedabad : Space Application Centre (ISRO).

3. "Economic Valuation of Tropical Wetland Resources : Report prepared for CATE and the Regional Wetlands Programme of IUCN, London : London Environmental Economics Centre (IIFD and University College, London), Application in Central America (1989 b).

4. Topographical Sheets No. 45O, 45P, 54C and 54D.

5. Claridge, G. (1991), "An overview of wetland 'Values' : A Necessary Preliminary to wise use', in R. Donoline and B. Phillips (eds) Educating and Managing for wetlands conservation and Management workshop, pp. – 119-132. Newcastle.

6. Trivedi, P.R., (1995); "Sustainable Management of Resources" (30 volumes), Indian Institute of Ecology and Environment, New Delhi.

Rural Land Use Management And Planning

An Optimal Stretgy for Commercial Farming in Kashmir Valley

S.A. Shah

Department of Geography and Regional Development,
University of Kashmir, Srinagar-190006

Introduction

Land is our ultimate asset. Apart from food and industrial crops, it provides employment to about 70 percent of the total population of the country. Land is however, finite and same is the case with arable land in all the countries and regions of the world. Since expansion of agriculture in the new areas demands heavy investment, especially in a mountainous region like the Kandi land of Kashmir Division, it is imperative to intensify agriculture by judicious utilization of land and by adopting suitable varieties of cereals and non-cereal crops. Agriculture is still the dominant economic activity in the Valley of Kashmir, generating about 50 per cent of the Gross Domestic Product of the Division and engaging nearly 70 percent of the workforce. Thus the economy of the region hinges on agriculture. The maintenance of resilience characteristics of the soils and making agriculture more productive and sustainable are the issues of vital importance which deserve special attention of the planners, decision makers and growers. The Valley of Kashmir, having a unique physical and geo-climatic settings is still tradition bound in agricultural practices. The physical environment and the socio-cultural milieu does not permit the farmers to diffuse new agricultural technology in their holdings. Consequently, the farmers are growing either paddy or maize and fodder crops. There are however, some lacustrine deposits, locally known as **Wudur** or Karewas which are ideally suitable for the cultivation of saffron. Saffron is a very expensive condiment which fetches handsome amount to its growers. It has been reported by the cultivators of saffron that the productivity per unit area of saffron is declining in the traditional saffron growing areas like Pampur.

Saffron, the leading cash crop of the Valley of Kashmir fetches handsome amount annually to its growers. Its cultivation and allied operations provide employment to about 15 percent of the rural work force. Moreover, it contributes about 25 percent of the total agricultural income of Kashmir Division. From the quality point of view, the Kashmiri saffron has a very high reputation in the International Market. Consequently, there is an increasing demand for this condiment. Kashmir produces less than half of the national consumption of saffron.

Production of Spanish saffron, which mostly originates in the central areas of the country, has been declining in recent years. If the import statistics are to be taken as indicative, it could be easily judged that there is great demand of Kashmiri Saffron in the international market. It is used in more than one ways as to give flavour and taste to food. It is also needed as a raw material in the pharmaceutical industry for the manufacturing of invigorating, life saving and curative medicines. The importance of saffron cultivation in the economy may be judged from the fact that its hectarage is often been taken as an indicator of the social status of the farmer.

Fig. 1

Objective And Scope

The study was undertaken with the aim of generating optimal land use plan and also study spatio-temporal variation in land use /land cover with special reference to a cash crop, Saffron (Crocus stivous) in Chandhara village, Pulwama District of Jammu and Kashmir state, India. . On the basis of micro-level study a suitable strategy to increase the productivity in the existing saffron growing villages will be suggested. Attempt has also been made to recommend the steps to be taken to make saffron as an economically viable crop in the unutilized Karewas of the Valley and to make its cultivation ecologically sustainable.

Study Area

The village CHANDHARA situated at 33^0 59 ' N latitude and at 74^0 56' E longitude, lies in Pampur Tehsil in the district of Pulwama. Situated about 16 km. to the South East of Srinagar

City end about 1.5 km. to the north of the Jammu-Srinagar National Highway No 1, the village is accessible by a metalled road. *CHANDHARA*, a comparatively prosperous village is famous because of a legendary poetess *Habba Khatoon (Zooni)*. It was in beautiful saffron dell of Chandhara that the then king, *Yusaf Shah Chak* saw Habba Khatoon for the first time and found her looks more intoxicating than the aroma of saffron. It was among these Saffron beds that the only *affaire decoeur* which was bold enough to rebel against the social norms, and made a king merry an ordinary folks women. One of those rare occasions in the history of world which saw the union of nobility and ordinary.

Data Base

The present study is based on primary and secondary data obtained from the published records and gathered from the saffron growing selected sample villages in the form of structured questionnaires, field interviews, observations and participation. Data regarding the general landuse and cropping patterns have been obtained from the **Patwaris** (land record keepers) of the concerned villages. The data for two term years 1994 and 2004 were used to see decadal spatio-temporal variations in the pattern of saffron cultivation. Some information was obtained from the Gazertteers, Census Reports, District Census Hand-books, published by the Town and Village Directory, Director of Census Operations, Jammu and Kashmir.

Methodology

Temporal data pertaining to Kharif (Rainy) and Rabi (Winter) season were used to delineate dominant agricultural crops and other land uses viz: Saffron, Almond, Paddy, Orchards, and Fallow etc. For temporal variations in crop acreage, data pertaining to different cropping seasons with a time interval of ten years was obtained for both seasons (Kharif and Rabi). The base map of the village in the form of cadastral map was obtained and digitized in GIS software viz: MapInfo. A database was created by feeding the thematic maps like land use/ land cover, soils, groundwater table and irrigation. The thematic maps for the two different cropping years (1994 &2004) were superimposed to study the decadal variation in land use pattern, which resulted in the land use/ Land cover change detection map.

Results And Discussion

Soil Classification

Fig. 2 shows spatial distribution of soils in village Chandhara. The soil of good quality land (Surzamin) which is rated as the best in fertility is devoted to vegetables. The Gurti-A, i.e., red Gurti (Karewa soil) is mostly devoted to saffron and almond crops. The other soils found in the village are Gurti-B or yellowish Gurti (Karewa soil) which is less fertile as compared to Gurti-A., is devoted to orchards (apple). The Bahil (loamy soil) is devoted to fodder crops. The Sekil (sandy soil) is covered by culturable waste. The Nambal

(Swampy & Clay soils) are devoted to paddy. It is highly productive soil for paddy. For ploughing and pulvarisation it however, demands hard work.

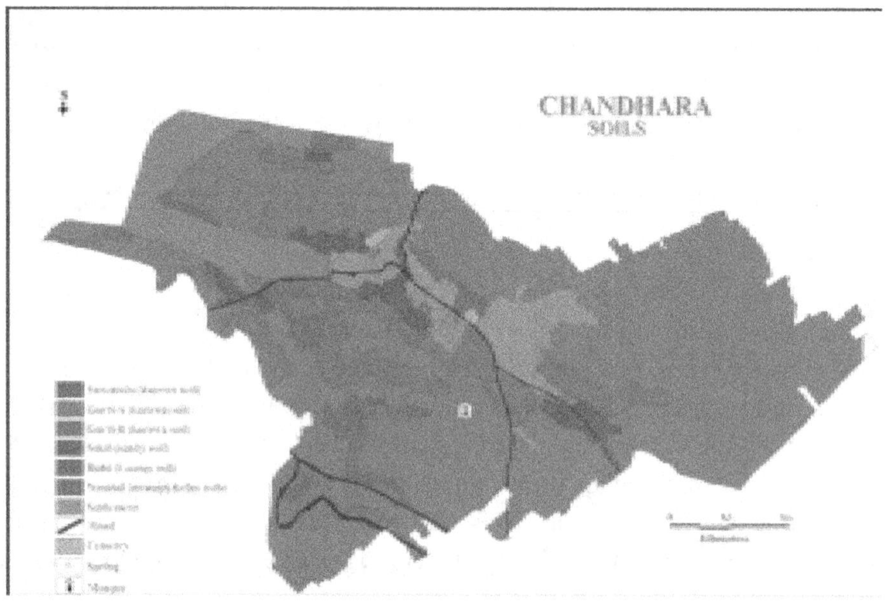

Fig. 2

Table1.1 Village Chandhara : Soils Classification

Soil	Classification	Area in acres	Percentage of the total area
Surzamin	(Karewa soil)	28.56	2.37
Gurti-A	(Karewa soil)	805.00	66.80
Gurti-B	(Karewa soil)	47.14	3.91
Sekil	(Sandy soil)	105.00	8.71
Bahil	(loamy soil)	9.5	0.78
Nambal	(Swampy & Clay soil)	100.00	8.29
Land put to non-agricultural use		109.00	9.11
TOTAL		**1205.00**	**100.00**

Source: Field Work by the author.

Table 1.1 shows that about 73% of the total area is covered by the Kerewa soils and about 8% by swampy soils. Sandy soils covers an area of 8% and lass than 1% is covered by loamy soils. The land put to non-agricultural use covers about 10% of the total area of the village. The most important aspect of the physiography of Chandhara is its location on a levelled tract of Flate Topped Karewas at the footsteps of famous summit *Wostarwan* 2965

MSL. metres. Its surface is characterized by an unvarying monotony of a levelled tract. The undulations are however, not absolutely absent as a large ravine traverses the village from one end to the other.

Table 1.2 General Landuse in Village Chandhara 1994-95
TOTAL AREA 1205 acres

Category of Land	Area in acres	Percentage of the total area
I Forests	00.00	00.00
II Land not available for cultivation	65.50	5.44
(a) Land put to non-agricultural use	65.50	5.44
(b) Barren and uncultivated land	00.00	00.00
III Other uncultivated land excluding Fallow	148.13	12.29
(a) Permanent pastures and grazing lands	00.00	00.00
(b) Miscellaneous tree crops and grove not included in net area sown	10.25	0.85
(c) Culturable waste	137.88	11.44
IV Fallow Lands	0.38	0.032
(a) Fallow Lands other than current fallow	00.00	00.00
(b) Current fallow	0.38	0.032
V Net area sown	991.00	82.24

Source : Field work by the author

It can be seen from table 1.2 that the total area of the village is 1205 acres (9640.00 Kanals). Out of this the percentage area under forests is negligible as the forests area have been brought under cultivation of crops. Land not available for cultivation covers about 65.50 acres (524 kanals) or 5.44 percent of the total area. Out of this land put to non-agricultural uses amounts to all the 65.50 acres and the barren and uncultivated land is non-existent in the village. Other uncultivated land excluding fallow share about 148.13 acres (1185.04 Kanals) or 12.29 percent of the total area of the village 'Permanent pastures and other grazing land' are nonexisting as the area has been brought under cultivation. 'Miscellaneous trees and other grove's account for 10.25 acre (0.85 percent) while 'culturable waste' is about 137.88 acres or 11.44 percent of the total area of the village. The culturable waste is confined to the slopes of the ravines and some water logged area. The common tree crops and groves include Keekar (Acacia-Arabia), willow, poplar, chinar, elm, mulberry and walnut. Fallow land covers about 0.38 acrea or 0.032 percent of the total area of the village. Due to increasing demand and population pressure very little area remains fallow. Every cultivable piece is being devoted to one or the other crop. Table further shows than only a small

139

proportion of land is sown more than once. The total cropped area thus becomes 991.00 acres; cultivable area comes about 1128.88 acres or 93.68 percent of the total village land. The possibility of bringing new area under cultivation is highly limited.

Table 1.3 Agricultural Landuse in Kharif Season 1994

Cultivated Land 991.00 acrea

Net cropped Land 990.99 acrea

Crops	Area in acres	Percentage of the gross cultivated Land	Percentage of the net cropped Land
Saffron	814.00	82.14	82.14
Almond	70.25	7.09	7.09
Orchards (Apple)	17.75	1.79	1.79
Paddy (rice)	73.00	7.37	7.37
Vegetables	16.00	1.61	1.61
Fallow	0.38	0.03	0.03
Total	**991.00**	**100.00**	**100.00**

Source : Field work by the author.

AGRICULTURAL LANDUSE IN KHARIF SEASON 1994

Chandhara shows a marked deviation in cropping pattern from the state of Jammu and Kashmir. In the state about 90.24 percent of the gross cropped area is under cereals, like wheat, rice and maize - while in Chandhara, about 82.14 percent of the gross cropped land is under saffron and less than 8 percent under cereals. Rest of the land is under Orchards viz; almond and apple. Table 1.3 Fig. 2.0 shows the various crops and their percentage share in the village Chandhara in Kharif season. A glance on the table reveals that the Kharif season occupies a dominant place in the agricultural cycle of the village like other parts of the Karewas in the Valley of Kashmir. It accounts for about 99 percent of the gross cropped area. Saffron is the dominant crop sown in the village. It accounts for about 82.14 percent of gross cropped land and 82.14 percent of net cropped area. Table 1.3 exhibits that an area of about 73.00 acres of land is devoted to paddy cultivation which accounts for 7.37 percent of the gross cropped land and 7.37 percent net cropped area. The crop has least potentiality of extension as the most of the area is of Karewa in which irrigation facilities could not be extended. Almost the entire area under paddy is in the western part of the village as it is situated on the lowlying area and can be irrigated from the Jehlum river through lift irrigation. At present most of the paddy area is irrigated by natural springs. It may also be observed from Table 1.3 that almond orchards covers an area of about 70.25

acres, 7.09 percent of net cropped area. Vegetables are of great importance in Kashmir Valley and every villager has small Kitchen garden where he raises a wealth of food with little efforts.[1] The common vegetables grown during Karif Season in the village are Knol-Kohl (Brassica Oleracea) or Krem Hak; tomato (Solanum Lycopersicum) or *Ruwongan*, Pumpkin (cucusbita pepo) or *Al*, bringal, *Olu* (Solnum Tuberosum) and *Cucumber* Cusumia sativus) or *Lar*. Besides these vegetable there are some condiments like Chilli (Capsium sp.) or *Mortswangan* and onion (Allavim sp.) or Pran also grown in these vegetable gardens. Among all Vegetables the *Knol-Kohl* a pot-herb hold a supreme position. It is regarded as a national vegetable for Kashmiris.[2] It is such a popular vegetable in the Valley that, on an average it is cooked at least thrice a week. An area of about 16.00 acrea or 1.61 percent of the net cropped area was devoted to kharif vegetable is the village. There is an area of about 17.75 acrea or 1.61 percent net area under apple orchards. About 0.38 acrea or 0.03 percent of the gross cultivated land was left fallow during the Kharif Season 1994.

Table 1.4 Agricultural Landuse in village Chandhara (2004-2005)

Total Area 1205 Acres

Category of Land	Area in acres	Percentage of the total area
I. Land not available for cultivation	109.87	9.12
(a) Land put to non-agricultural use	109.87	9.12
(b) Barren and uncultivated land	0.00	0.00
II. Other uncultivated land excluding fallow	133.25	11.06
(a) Permanent pastures and grazing lands	0.00	0.00
(b) Miscellenous tree crops and grove lands not included in net area sown	8.50	0.70
(c) Culturable waste	124.75	10.35
III. Fallow Lands	61.25	5.08
(a) Fallow land other than current fallow	0.00	0.00
(b) Current fallow	61.25	5.08
IV. Net area sown	900.62	74.74

Source : Field work by the author.

It can be seen from table 1.4 that the total area of the village is 1205 acres. Out of this not even a single acre is under forest. 'Land not available for cultivation covers about

109.87 acres (878.96 Kanals) or 9.12 percent of the total area. Out of the land put to non-agricultural uses amount to all the 109.87 acres. There is no barren land in the village. 'Other uncultivated lands excluding fallow share about 133.25 acres (1066.00 Kanals) or 11.06 percent of the total area of the village. 'Permanent pastures and grazing lands are nonexistent. Miscellaneous trees and grove lands account for 8.50 acres (68.00 kanals) or 0.70% of total area of the village. Culturable waste accounts for 124.75 acres (998.00 Kanals) or 10.35% of the total area of the village. Fallow lands covers about 61.25 acres (490.00 Kanals) or 5.08% of the total area of the village. The net sown area comes about 900.00 acres (7204.98 Kanals) or 74.74% of the total area of the village.

Table 1.5 Agricultural Landuse in Kharif Season 2004

Cultivated Land	961.87	Net Cropped Land	900.62
Crops	**Area in acres**	**Percentage of gross cultivated cropped land**	**Percentage of the net land**
Saffron	659.38	68.55	73.21
Almond	59.00	6.13	6.55
Orchards (Apple)	8.125	0.84	0.90
Almond & Saffron	68.37	7.11	7.59
Paddy (rice)	39.87	4.15	4.43
Vegetables	28.50	2.96	3.16
Fodder	8.63	0.89	0.96
Maize	28.75	2.99	3.19
Fallow	61.25	6.37	-
Total	961.87	100.00	100.00

Source : Field work by the author

It can be vividly seen from Table 1.5 Fig. 4.0 that in 2004 the area under saffron is 659.38 acres which accounts for 68.55 percent of gross cropped and 73.21 percent of net cropped area. A new trend has come into practice i.e., double cropping of almond and saffron are being cultivated together. It accounts for 68.37 acres, 7.10 percent of gross cultivated land and 7.59 percent of net cropped land. The area under almond 59.00 acres or 6.13 percent of gross cropped area and 6.55 percent of net cropped area. Maize is the cereal crop which has been introduced in recent years and accounts for 28.75 acres or 2.99 percent and 3.19 percent gross and net cropped respectively. Fodder crop (cutting of maize before it bore seed) is other important feature in cropping pattern of Kharif season 1994. Fodder crops covers an area of 28.75 acres which accounts 2.99 percent of gross cultivated area and 3.19% net cropped land. Paddy (rice) and orchards (apple) which was cultivated before is still having its portion in Kharif 1994, it was cultivated on an area of 39.87 acres and 8.12 acres respectively. The area under vegetables is 28.5 acres or 2.96 percent and 3.19 percent of gross cultivated and net cropped area respectively. The fallow area amounts 61.25 acres and 6.37 percent of gross cropped area.

Comparative Study Of Landuse

In order to ascertain the variations in landuse and cropping pattern, it is imperative to make a comparative analysis of landuse scenario of the village between 1994-2004. Such a comparison will help in ascertaining the changes that have taken place in the landuse patterns of the village during the last ten years. A comparison of general landuse in 1994-95 and 2004-05 shows a slight decrease in the net area sown from 991 acres to 900.62 acres, and increase in the fallow land from 0.38 acres to 61.25 acres. Where approximately 61.00 acres of the net sown area fell fallow, the rest of the decrease in sown area was due to new settlements which shows an increase from 65.50 acres to 109.87 acres. Slight portions from uncultivated land and cultivable waste came under settlements.

A cross examination of landuse in Kharif 1994 and 2004 reveals that saffron cultivation was spread over an area of 814.00 acres which amounted 82.19 percent of net cropped area during Kharif 1994, while in Kharif 2004, it decreased to 659.38 acres which amounts 73.21 percent of net cropped area and 65.88 percent of gross cropped area. This decrease of 9 percent approximately in net cropped land and 14 percent approximately of gross cropped land may be correlated with the declining fertility of soil owing to over exploitation of its cover for generations. This trend of decrease is quite significant in the case of almond, orchards and paddy cultivation, which decreased during this decade from 70.25 to 59.00 acres, from 17.75 to 8.13, and from 73.00 to 39.87 acres respectively. The decrease being highest in the area under paddy cultivation and saffron cultivation. An increase in the land under vegetables occurred during this decade from 16.00 to 28.30 acres. Fodder and maize which were not cultivated during 1994 went under cultivation in 2004 Kharif season. Although the changes in the area under various crops are significant and seem to be owing to the diffusion of innovations, agriculture, the decrease in area under saffron cultivation, may be attributed to the declining per acre yield and theuncertain price policy. It has been reported by the villagers that the productivity of soil under continuous saffron cultivation had dwindled. The decrease in land under paddy cultivation during 2004 was according to farmers mainly because of the water logging of areas which grow paddy. This decrease was only in the concerned year and does not reflect any decreasing trend in the preference for this particular crop which continues to supply the village with their staple food. However the increase in vegetable cultivation and land under fodder crops is mainly because of market influences. The rates of the vegetables increasing day by day compels the villagers to be self sufficient for which the farmers cultivate vegetables on more and more land. The non-availability/high prices of the fodder, hand in hand with increase in number of livestock in the village Chandhara that is substantiated by the fact that the village was not self sufficient in milk during eighties, now-supplies about 700 litre milk daily to Srinagar city, compled the farmers to bring more land under fodder crops.

Suggestions

Some of the steps which may go a long way in making saffron cultivation more productive are as under:

1. Over the centuries saffron is being grown by traditional techniques, resulting into depletion of soil fertility. A scientific rotation and diffusion of new agricultural innovations are imperative to enhance area and production.

2. Inter cropping of almond and saffron in the Karewa lands is on the increase. The large trees of almond are however, not good for better saffron yeilds as saffron does not thrive well in the trees of almonds as their large roots create obstacles for the tender corms of saffron. Therefore development of new varieties of almond trees is very essential which may increase the yield per unit area of both these commercial crops.

3. Under the growing pressure of population many of the good saffron fields are being brought under houses and settlements. This practice needs to be stopped for which enforcement of Saffron Act is to be implemented vigorously.

4. Research Stations should be located at different saffron growing Karewas rather to keep them restricted to the lands of a few elite farmers and politicians.

5. Fragmentation of holdings is a serious setback in the efficient management of the crop. The government should initiate the process of consolidation of holdings without further delay. The consolidated holdings can help in the fencing of the fields which shall ultimately leads to better management of the crop.

If all the steps suggested above are collectively taken, the production of saffron will go up significantly and the fertility of the soil will be maintained. In other words saffron cultivation will not only be economically more rewarding, it will be ecologically more sound and sustainable

References

1. Lawerence, W. R. 1961 , *The Valley of Kashmir*

2. Kumar , P. 1990 , *Saffron Story Indian Spices* , 27 (1) : 5-11

3. Hussain, M. !996 , *Systematic Agriculture Geography* , Rawat Jaipur.

4. Mir, G. M. , 1983, *Saffron Agronomy in Kashmir*, Gulshan Publishers, Sgr.

5. Stamp, L. D., " The Measurement of Land Resources", *Geographical Review* , 48 (1958) 1-9.

Wastelands Information System for Sustainable Development

R. Nagaraja , Rajiv Kumar, B. Shyamsunder, Manoj Raj Saxena, Satish Jayanthi

Land Use Division, National Remote Sensing Agency, Hyderabad – 500 037

ABSTRACT

The Indian experience, in use of remotely sensed data for Land Use / Land Cover Analysis, gained over past more than 20 years of implementation of various projects especially "Nation-wide Land Use / Land Cover Mapping for Agro Climatic Zone Planning" and 'National Wasteland Inventory Project' are described. The former project was sponsored by Planning Commission of India and later by Ministry of Rural Development (MRD) of Govt. of India. A systematic study was carried out to identify and map 13 different types of wastelands on 1:50,000 scale upto village and micro watershed level. A digital data with standard codification system in four different layers were generated for the entire country. About 64 million ha have been estimated as wastelands through this study. Various watershed programmes are being implemented in the country consulting this database. A successor to this project "Monitoring of Wastelands", which aim at reporting the changes in wastelands statistics in India using IRS LISS III 2003 data, considering various developmental programmes launched by Govt. of India since early 1990 has been completed. The latest study revels that the extent of wastelands as 55.27 M. ha., a reduction of 8.58 M.ha.(2.71 per cent) Realising the importance of spatial land use / cover information and it change pattern for planning and developmental activities, a land use census mission, under Natural Resource (NR) Census, which aims at generation of land use / cover information including wasteland classes at multiple national scales with synergistic use of all information sources is expected to be soon implemented under National Natural Resource Information System (NNRMS) programme.

Introduction

Information on land use / land cover in the form of maps and statistical data is very vital for spatial planning, management and utilisation of land for agriculture, forestry, pasture, urban-industrial, environmental studies, economic production etc. Today, with the growing population pressure, low man-land ratio and increasing land degradation, the need for optimum utilisation of land assumes much greater relevance. The draft outline on the National Land Use Policy and strategy on Optimum Land Use Planning and the creation of National Land Use Conservation Board (NLUCB) in 1985 clearly indicate the serious concern of the Government in this regard. Further, with the present thrust that the agricultural planning in the country should be based on Agro-Climatic Zones, the prima-facie need is to have a comprehensive information on the spatial distribution pattern of land use / land cover, particularly on the availability of agricultural land during Kharif and Rabi crop seasons, cropped area during both seasons and area under fallow, apart from the other land use / land cover classes. Besides, it is also required to know their area for the whole country covering all the 15 agro-climatic zones in the 442 districts (now the number of districts are 584). This needs land use inventory surveys periodically, to make available the information on the type, spatial distribution, location, areal extent, rate and pattern of change of each category of land use / land cover on the land. Preparation of an upto date, accurate and reliable information on land use / land cover over large areas on a contiguous basis is possible using remote sensing techniques on an operational, timely and on a cost effective basis.

Today, the availability of information on land use / land cover in the form of thematic maps, records and statistical figures are inadequate and do not provide an upto date information on the changing land use patterns and processes. Over the years, the efforts made by the various Central / State Government Departments, Institution / Organizations etc., is sporadic and often efforts are duplicated. In most the cases, as the time gap between reporting, collection and availability of data is more, the data often becomes out-dated. However, the organisational efforts in publishing maps, reports and statistical data by the Survey of India (SOI), National Atlas & Thematic Mapping Organization (NATMO), National Bureau of Soil Survey & Land Use Planning (NBSS & LUP), All India Soil & Land Use Survey (AIS & LUS), Central Arid Zone Research Institute (CAZRI), Ministry of Agriculture, Settlement Survey and Land Records, Revenue Department, National Sample Survey, State Land Use Boards, Town & Country Planning Organization (TCPO) and other local agencies are noteworthy.

National Wastelands Inventory Project (NWIP)

Until recently, no attempt had been made to prepare a comprehensive map on any scale showing the distribution of different types of wastelands in India. A number of organizations and agencies commenced the collection and collation of data on the type and extent of wastelands. The area reported by various government agencies are varies from 38 M Ha. To 175M. Ha. The considerable variation in the area underscores the need to prepare a reliable database using the latest techniques of satellite remote sensing and geomatics.

In 1985, the National Remote Sensing Agency (NRSA) of the Department of Space prepared wasteland maps of all the Indian states and union territories at a 1:1 million scale. The total area of wastelands in 1980 – 1982 estimated through this study was about 53.3 million ha or 16.2 per cent of the area of the country.

The 1:1 million scale wasteland maps of the states and union territories of India provided only a gross estimate of wasteland area. Due to the small scale and relatively poor resolution of the MSS imagery, wasteland areas of less than 100 ha could not be mapped. Further more, it was found that the maps were of little use in reclamation planning at the local level.

In 1985, the then Prime Minister of India set up a National Wastelands Development Board (NWDB) with the objective of rehabilitating 5 million ha of land each year for fuel wood and fodder production through a massive programme of seeding and afforestation. This programme required a very reliable database that provided details on the type, extent, location and ownership of wastelands.

Wastelands in India as per 1986-2000 Satellite Data

Methodology developed based on pilot studies were used for the identification and delineation of different types of wasteland using enlarged satellite data. Both Landsat Thematic Mapper (TM) and Indian satellite (LISS-II and LISS-III) data were used for mapping purposes. Hybrid methodology i.e. both visual and digital techniques were used to extract the wastelands thematic details.

Village boundaries, forest compartment and watershed (upto micro watershed having 500 ha area) boundaries were incorporated into the wasteland maps. The final product is a wasteland map with village, forest compartment and micro watershed boundaries at 1:50,000 scale. All these information is available in the digital form covering 8000 + map sheets on 1:50,000 scale. It is possible to associate wasteland units with particular villages / micro watersheds using these maps. In addition, possible ownership can be determined by comparison with cadastral maps available at scales of 1:4000 and 1:8000. The vector based digital database generated under this project using standard codification was organized in four different layers viz base layer, administrative layer, watershed layer and wasteland layer.

About 63.87 million ha (20.17 per cent) have been estimated as wastelands through this study. Category-wise and State-wise area under wastelands are shown in tables 6 & 7. About 5000 wasteland maps covering the country were prepared under this project. After completion, the maps are sent to various state users for wasteland reclamation measures. Brief reports for each district provide a general description of the wasteland and its distribution.

User interaction workshops have been held in different states to demonstrate the utility of wasteland maps in reclamation activities and to instruct users in making the best use of such materials. An atlas covering the statistics of entire country is released for public by hon'ble Prime Minister of India Sri A. B. Vajpayee on 22nd May, 2000 in New Delhi.

All these 63.87 million ha are highly degraded lands in the country. Apart from this, the process of degradation exist in the agricultural lands such as single cropped areas and fallow lands. Based on the statistics generated under land use / land cover project and the wasteland inventory project, it is estimated, about 145 million ha are under degraded land i.e. 63.85 M ha under severe extreme degraded lands, 13.76 M ha under moderately degraded lands and 67.49 Mha under slight to moderate degraded lands as shown below.

Degraded Lands in India

CATEGORY	AREA (M HA)
SEVERE & EXTREME (WASTELANDS)	63.85
	13.76
MODERATE (FALLOW LANDS)	67.49
SLIGHT TO MODERATE (SINGLE CROP-KHARIF FALLOW)	
TOTAL	145.10

ABC Categorization of Wastelands

Since there is financial and time constraints, it is obviously not possible to develop all the wastelands / degraded lands in a short period. Therefore, it is essential that the Department may concentrate on certain wastelands which require aggressive intervention through Government funds for development.

Keeping this in view, a road map needs to be designed with properly marked off milestones and sign posts in the field of watershed / wasteland management.

For suggesting a proper and focused management the wastelands / degraded lands were classification into following ABC categories considering the priorities and other developmental strategies.

A - Those lands which requires high capital, modern technology and long term implementation strategies. These lands are like cold snow clad deserts, hot desert etc.

B -   Those lands which require direct Government Action – predominantly in the form of subsidy, grant etc., and aim at environmental improvement on community basis. In this category backward forward linkages etc., are required.

This "B" category can be further sub-divided into 1, 2 & 3 categories.

B1 category is one where governmental intervention becomes a must on account of the poor response, low absorption capacity etc. These land need special attention the role of the other players are limited and they do not have a meaningful role in this field.

B2 category is for those lands where some infrastructure is development. There is a potential for response and where some sort of linkages, awareness etc., are already in existence and therefore these lands can be developed by inducing catalytic action arranging availability of inputs by involving other players and putting in place appropriate backward and forward linkages with asset creation and individual livelihood systems being brought to center stage.

B3 categories are those where proper managerial strategies are required.

In all these B1, B2 and B3 categories a further classification of V, E, D i.e., vital, essential and desirable categorization is to be done.

C - Those lands which are on the threshold of a new economic order at a local level where resolution of second generation issues and addressing sustainability concerns are required.

The database generated under the NWIP project were sub categorized as per the above requirement. Government of India, MRD has used this data base for implementation of watershed / wastelands development programmes in India, through government departments, NGO's, co-operative societies etc. To assess the impact of these programmes wastelands monitoring using satellite data (Feb/ March 2003) has been taken up & completed.

Wasteland Classification

Confronted by varying estimates of the extent of wasteland, including the NRSA figure based on remote sensing, it became evident that the NWDB had to provide precise definitions of the various categories of wasteland. The Technical Task Force established by the NWDB proposed a classification system consisting of thirteen categories of wastelands, and later modified to 28 sub classes to map the degree of degradation as indicated below.

WASTELAND CLASSIFICATION SYSTEM

1.	Gullied and/or Ravinous land	1.1	Shallow
		1.2	Medium
		1.3	Deep/Very Deep
2.	Land with or without Scrub	2.1	Land with Scrub
		2.2	Land without Scrub
3.	Waterlogged and Marshy land	3.1	Permanent
		3.2	Seasonal
4.	Land affected by Salinity/Alkalinity	4.1	Strong
		4.2	Moderate
		4.3	Slight
5.	Shifting Cultivation	5.1	Current Jhum
		5.2	Abandoned Jhum
6.	Under Utilised/Degraded Notified Forest Land	6.1	Scrub domenated
		6.2	Agricultural land inside notified forest land
7.	Degraded Pastures/Grazing Land	7.1	Degraded Pastures/Grazing Land
8.	Degraded Land Under Plantation Crop	8.1	Degraded Land Under Plantation Crop
9.	Sands (Riverine/Coastal/Desert)	9.1	Flood Plain
		9.2	Levees
		9.3	Coastal Sand
		9.4	Semi-Stabilised to Stabilised (> 40 m)
		9.5	Semi-Stabilised to Stabilised Moderately High (15-40 M) Dune
		9.6	Semi-Stabilised to Stabilised Low (< 15 M)
		9.7	Closely Spaced Inter Dunal area
10.	Mining/Industrial Wasteland	10.1	Mining Dumps
		10.2	Industrial Wasteland
11.	Barren Rocky Area	11.	Barren Rocky Area
12.	Steep Sloping Area	12.	Stony Sloping Area
13.	Snow Covered and/or Glacial Area	13.	Snow Covered and/or Glacial Area

Distribution of wastelands in India as per 2003 data

About 55.27 million ha. forming 17.45 per cent of the wastelands to the total geographical area of the districts covered, was estimated through this exercise. State wise and category wise distribution of wastelands are shown in the tables. A sample map of Wasteland along with base, watershed and administrative layers of Goa state is shown in the figure.

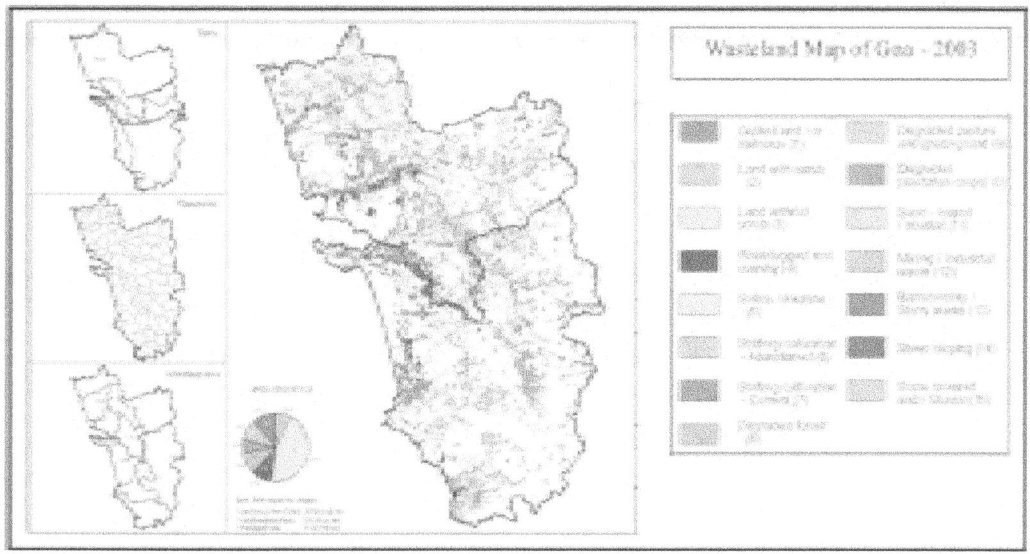

The very high percentage of area under wasteland in Jammu and Kashmir (69.24%), Himachal Pradesh (50.90%), are due to snow cover and degraded forest; Nagaland (22.37%), Manipur (59.01%), Mizoram (21.20%) are due to shifting cultivation, Sikkim (53.67%) is due to degraded forest and in Rajasthan (29.64%) due to Sandy area. Among all the states Punjab has a minimum 2.33 percent and Jammu & Kashmir has a maximum 69.24 percent of area under wastelands. The category-wise distribution of wastelands shows that highest percentage (4.76%) belongs to the category 'land with scrub' followed by 'under utilised forestland' (3.42%). The former is mainly distributed in the southern states of India whereas the later is distributed throughout the country. The total Geographic area of the country is covered in 597 districts, out of which fourteen (14) districts are newly formed and the statistics are not generated separately as these are covered as combined district. The wasteland statistics for these fourteen districts can be extracted after obtaining authentic boundaries. In view of this, analysis carried out for 583 districts for which wasteland statistics are generated. Spatial location of the districts with state-wise district code are shown in plates. Among the 583 districts, 19 districts mainly distributed in NE states, Jammu & Kashmir and Rajasthan have got more that 50% of geographical area under wastelands. 103 districts have 41-50%, 76 districts have 31-40%, 98 districts have 21-30%, 126 districts have 11-20% & 151 bdistricts have less than 10%. Ten districts are having more than 5,00,000 ha mainly distributed in J&K, Rajasthan and Gujarat, 49 districts between 2,00,000 to 5,00,000 ha, 89 districts between 1,00,000 ha, 114 districts

Wasteland dynamics

Wasteland mapping was carried out for two time periods ie. 1986-2000 and 2003. The first cycle was carried out in five phases using three period data sets i.e. 1986-88, 1991-92 and

1997-98. The second cycle was completed using one year i.e. 2003 data. The comparative analysis of statistics shows that there is reduction of 8.58 million hectares area between those time periods, equivalent to 2.71% to total geographical area of the country. In three states namely, Meghalaya, Nagaland and Gujarat, the reduction in wasteland area is more than 10%, can be due to shift from shifting cultivation to permanent cultivation in the North Eastern part of states and due to reduction/elimination of Rann of Kutch in Gujarat. In another 19 states, the reduction in wastelands vary between 0.42% to 7.63%, due to various watershed programmes implemented. Part of the changes could be due to the methods and classification adapted of the analysis, data sets used etc. There is an increase in wastelands in some states mainly due to change in snow cover in states like Jammu & Kashmir, Sikkim and Uttaranchal. There is an increase in wastelands in Kerala by 0.88% mainly due to inclusion of new category namely, seasonal waterlogged area which is not mapped in first cycle. Category wise change analysis indicates that there is a reduction in wasteland areas in all the categories except for mining / industrial waste and steep sloping area. As an example changes in the wastelands in a watersheds of Nellore district Andhra Pradesh is shown below.

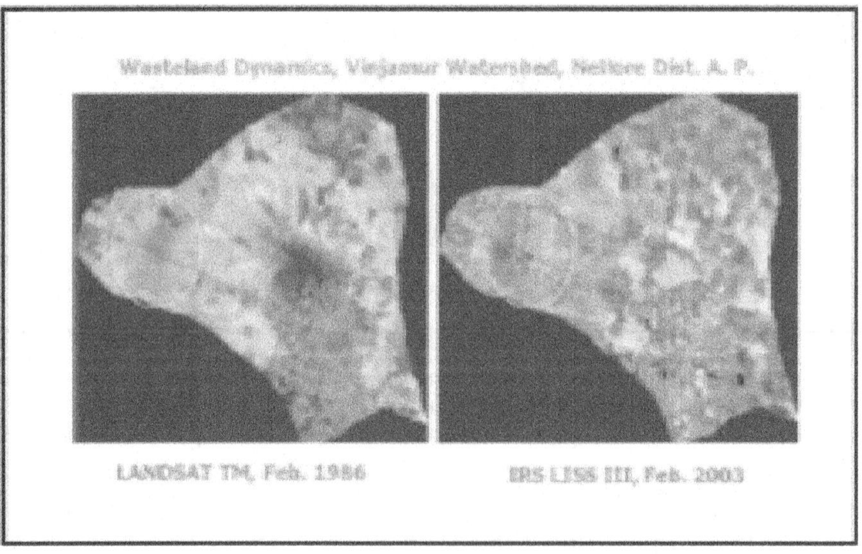

Conclusions

Recent advances in satellite sensor spectral, spatial and radiometric capabilities have strengthened the operational scenario of remote sensed based land use / land cover analysis on different scales / levels. Land use cover change (LUCC) information at National and global scale is also very important for modeling the global environmental changes. Proposed sensors on board ISRO's new satellites to be launched in next two years such as a WiFS (60 m, 4 visible – NIR Channels and 5 day repeat cycle) and LISS-IV (5.8 m pixel, 3 visible NIR channels) on Resourcesat will substantially improve the capability to generate LUCC information at Nation, regional and local scales. Availability of population and other socio-economic data at village level, every 10 year interval, will enable to integrate the LUCC information with socio-economic data for better understanding the driving forces for land use changes.

Acknowledgements

The authors are thankful to all the team members of NWIP and National Wasteland Updating Mission project from NRSA and different collaborating agencies for their contribution to the project and sharing of results and experiences that are described in this paper.

References

1. National Remote Sensing Agency (NRSA) (1985), Survey Report on Wasteland Mapping of India, Hyderabad, pp 1 – 10.

2. NRSA (1989), Manual of Nationwide Land Use / Land Cover Mapping using Visual Interpretation Techniques, Part – 1, Hyderabad pp 1-58.

3.   NRSA (1992), A Report on Reconciliation of Land Use / Land Cover Statistics generated by Remote Sensing and Ground Based Techniques, Hyderabad pp 1- 80.

4.   NRSA (1995), All India Report on Area Statistics of Land Use / Land Cover generated Using Remote Sensing Techniques, Hyderabad pp 1-71.

5.   Government of India, Ministry of Rural Development and National Remote Sensing Agency (2000) Wastelands Atlas of India.

6.   Gautam N.C., Ravi Shankar G., Narasimha Rao K, Nagaraja R, Manoj Raj Saxena, Jayanthi S.C. & Suresh L. S. (2000): Statistical Analysis of Land Use / Land Cover over India using satellite based Remote Sensing Techniques. Indian journal of Agricultural Economic Vol. 55, No. 2, April – June 2000.

7.   Nagaraja R, Gautam N. C., (1996) The role of Remote Sensing and GIS in Wasteland Management in India, Land Degradation in the Tropics – Environmental and Policy Issues, Published common wealth foundation.

8.   Food and Agricultural Organization (FAO) (2000) Land Cover Classification System: Classification concepts and use manual.

9.   NRSA (2000), Wastelands Atlas of India – 1866-2000.

10.   NRSA (2005), Wastelands Atlas of India – 2003

Watershed Management

Use of Remote Sensing and GIS in Watershed Management

B. R. M. Rao and M.A.Fyzee
Soils Division
National Remote Sensing Agency, Hyderabad - 500037

ABSTRACT

Among the contemporary technologies remote sensing and GIS technologies are proved to be valuable tools in characterizing the watersheds in terms of their natural resources, creation of data base and integrated analysis of information obtained through convention al and remote sensing methods to arrive at optimum solutions for various problems existing in the watersheds. Watershed level planning requires a host of inter-related information to be generated and studied in relation to each other. Geographical Information System (GIS) is a very powerful tool for development of the watershed area with all natural and socio-economic facets for better planning, execution and monitoring of the project. Its capability of integration and analysis of spatial, aspatial, multi-layered information obtained in a wide variety of formats both from remote sensing and other conventional sources has proved to be an effective tool in planning for watershed development.

A Study was undertaken using Remote sensing and GIS techniques in watershed management under NATP, an ICAR sponsored project which envisages natural resource inventory of watersheds, development of methodology for identification of critical areas and prioritization, detailed resource inventory of selected micro watershed, development and implementation of action plans and monitoring the impact through RS &GIS techniques. The project also envisages close interaction between planner and farmer. The results of work carried out in one of the watersheds are presented in this paper.

In this project action items such as construction of water harvesting structures, soil conservation measures, crop improvement techniques, etc were identified and implementation of the plans was carried out in the micro watershed. The study also showed how the Remote sensing technology can also help in monitoring the treated watersheds.

Introduction

Remote sensing has provided a new impetus for the earth resource and environmental scientists. This technology of space has to be fully harnessed for tackling the problems of the country. Synoptic view and quick decision-making process have become essential to meet the challenges appearing over the national scene. At this juncture, remote sensing technology has come to the rescue.

Remote sensing applications in the country, cover diverse fields such as soils, land degradation, crop acreage and yield estimation, drought assessment and warning, flood control and damage assessment, land use/land cover information, management of watersheds and command areas, agro-climatic planning, wasteland management, water resources management, underground water exploration, prediction of snow-melt run-off, fisheries development, mineral prospecting, forest resources survey etc. and now, with the advent of high resolution satellites new applications in the areas of urban sprawl, micro watershed development, infrastructure planning and other large scale applications of mapping have been initiated.

Watershed Approach

The 'watershed approach' represents the principal vehicle for transfer of rainfed agricultural technology. A watershed (or catchment) is a geographic area that drains to a common point, makes it an ideal planning unit for conservation of soil and water. It enables a holistic development of agriculture and allied activities. This systems-based approach is the special feature that distinguishes watershed development from earlier plot/field-based approaches to soil and water management. The key attributes of watershed management are conservation of rainwater and optimisation of soil and water resources in a sustainable and cost effective mode. It aims to optimise moisture retention and reduce soil erosion, thus maximising productivity and minimising land degradation. Improved moisture management increases the productivity of improved seeds and fertiliser, so conservation and productivity enhancing measures become complementary. Currently a large number of projects for productivity enhancement are being implemented based on the watershed approach.

Watershed Management

- The amount and nature of precipitation is the most important factor which determines what will happen in a watershed. Rainfall evenly dispersed throughout the year has a different impact

- from sudden. sharp showers or seasonal rainfall

- The drainage pauern of an area refers to the design of the stream courses and their tributaries. It is influenced by the slope of the land. lithology and structure.

Watershed management is the rational utilization of land and water resources for optimum production with minimum hazard to natural resources. It essentially relates water to soil and water conservation in the watershed which means proper land use, protecting land against all forms of deterioration, building and maintaining soil fertility, conserving water for farm use, proper management of local water for drainage, flood protection and sediment reduction and increasing productivity from all land use.

Different objectives call for different techniques, manpower, inputs and approaches in planning. The monitoring and evaluation criteria will also be different. Therefore, the main objectives should be identified and defined considering the priority and pressing need. Some of the common objectives can be:

- To rehabilitate the watershed through proper land use and to undertake protection / conservation measures in order to minimize erosion and simultaneously increase the productivity of the land and income of the farmers.
- To manage the watershed in order to minimize natural disasters such as flood, drought and landslides etc.
- To protect, improve or manage the watershed for the benefit of water resources development (domestic water supply, irrigation, hydro-power etc.)
- To develop rural areas in the watershed for the benefit of the people and the economies of the region.
- A combination of above.

Watershed Management Plan

A watershed management plan is essentially rural development programme and to derive the optimum benefits. Its approach should be multidisciplinary, involving aspects of cultivated crops, grasses, forestry, conservation engineering, animal husbandry etc. The sectoral plan may include adoption of simple traditional measures, like bunding, bench terracing, afforestation etc. It also includes a large number of moisture conservation structures like check dams and other structures including ponds, which come into use for multiple purposes such as control of erosion, prevention of encroachment of gullies into productive land, harvesting and storing of water for re-use, reclamation of degraded land etc.

Role Remote Sensing and GIS in Watershed Management

Natural resource survey of the area is the first step in any watershed development plan. It is done to assess the problems and prospects of the area. Such surveys carried out through traditional / conventional methods are tedious and time consuming. They can lead to human bias and are expensive. Moreover, remote sensing due to its inherent advantages has become a powerful a tool for natural resource survey and management. Its use for faster assessment of natural resources such as soil, geology, drainage etc. as well as assessment of economic activities through land use and infrastructure of the watershed area is well known.

Further, watershed level planning requires a host of inter-related information to be generated and studied in relation to each other. Remotely sensed data provides valuable and up-to-date spatial information on natural resources and physical terrain parameters. Geographical Information System (GIS) is a very powerful tool for development of the watershed area with all natural and socio-economic facets for better planning, execution and monitoring of the project. Its capability of integration and analysis of spatial, aspatial, multi-layered information obtained in a wide variety of formats both from remote sensing and other conventional sources has proved to be an effective tool in planning for watershed development.

The survey of literature reveals these techniques are employed in addressing various aspects of watersheds at 1: 50,000 scale or larger(Rao and Venkataratnam 1992, Rao *etal* 1998, NRSA, 1996, Saxena *etal, 2000,* Rao *etal* 2004, The utility of GIS has been used in the development of digital data bases, assessment of status and trends of the resources of the areas and to support and assess various resource management alternatives (Clark , 1990)

From 1996 onwards, with the availability of high spatial resolution satellite data from PAN (5m) and multispectral LlSS-111sensors (24m) from IRS - 1C and 10 satellites, provided an opportunity to study the natural resources at larger than 1:50,000 scale. A hybrid data product from merging of data from PAN and LlSS-111sensors, enable to generate FCC (False color composite) product at 1:12,500 scale and helps to generate various thematic maps. However, a very limited work has been reported on a microwatershed level by using high resolution satellite data. Subsequently, IKONOS and Quick Bird satellites were launched with sensors that can provide PAN data with 1.0m and 0.6 m spatial resolution, and multi-spectral data in 4m and 2.5 m. These data sets can be used to prepare the resources maps at 1:8,000 scale with multispectral data alone and at 1:4,000 scale by combining PAN and multispectral data sets. The utilisation of these data sets in microwatershed study has just began and it has to go a long way before they are used on an operational mode. In 2003, India has launched IRS-P6 (Resourcesat) with high resolutions advanced sensors viz., LlSS-IV (5.8m spatial resolution), LlSS-III (SWIR band 24 m spatial resolution) and AWiFS (56m spatial resolution with 10 bit radiometric resolution) and the satellite can provide data to study the natural resources from watershed to micro-watershed level.

Major areas where remote sensing and GIS techniques plays a vital role in watershed development are as follows:

1. Watershed delineation and codification
2. Watershed characterization

3. Watershed prioritization
4. Watershed planning and implementation
5. Watershed monitoring and impact assessment

Recently, a project under National Agricultural Technological Project (NATP) entitled "Development of regional scale watershed plans for identification of critical areas for prioritized land treatment in the watersheds of rice, oilseeds, pulses, cotton and NCRL production systems" was taken up by NRSA in association with ICAR collaborating centres. All the natural resources such as soils, ground water and land use /land cover were mapped in the selected watersheds at 1: 50000 scale using IRS IC /ID LISS –III data.

A methodology was developed for identification of critical areas based on soil parameters. Initially, the soil map of the watershed was prepared at 1:50,000 scale using latest satellite data from IRS – IC / ID satellites following visual interpretation approach with soil profile study, soil analytical data and classification of soils as per Soil Taxonomy. Digital database for the soil map was prepared using ARC / INFO GIS package and attribute data were incorporated for each soil-mapping unit. The parameters considered for identification of critical areas in the watershed are soil depth, soil texture, internal drainage, slope, erosion and soil reaction. The soil polygons were reclassified based on the above parameters and individual thematic layers were generated (Fig. 3) Area statistics were generated for the watershed parameter-wise to identify the critical parameter in the watershed from crop productivity point of view. A composite layer was generated from the individual thematic layers and the critical areas were mapped based on the limitations of the above mentioned soil parameters.

Rules were framed to group polygons with severe limitations under each parameter as high priority zone and polygon with no limitations under least priority zone. Polygons with moderate limitations were grouped under medium priority zone. This knowledge base was applied on to the composite layer and composite polygons were reclassified and map showing priority zones was developed

After characterizing the above watersheds at 1:50,000 scale action plans were generated for the development of land and water resources. For the implementation of action plans over such large areas involves huge financial and human resources. Therefore, critical areas for priority treatment were identified. Critical area map of Katepurna watershed, Akola Dt. Maharastra is shown in Fig.1.

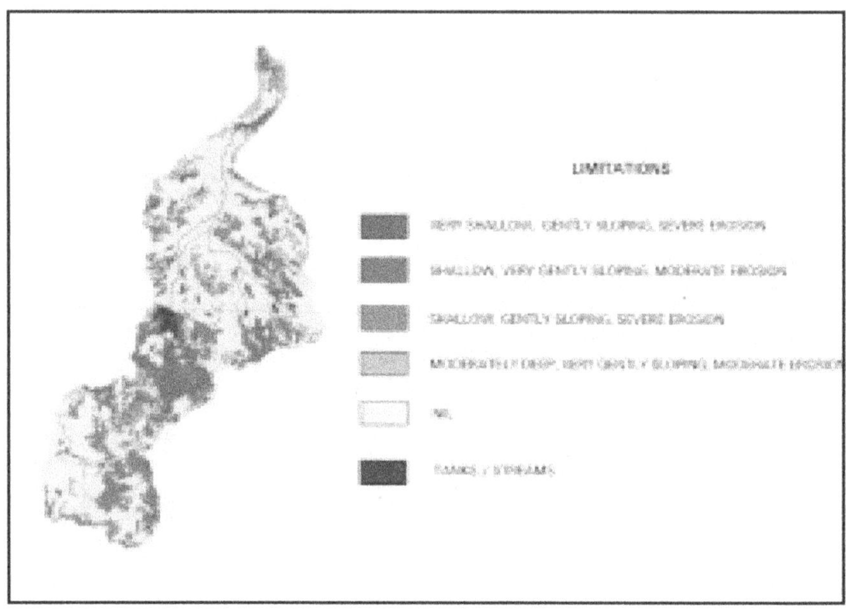

Fig.1 Critical areas of Katepurna Watershed, Akola

A representative micro watershed of an area of 500 – 1000 ha was selected from the critical areas for preparing natural resource inventory at 1:12,500 scale using high resolution satellite data from PAN & LISS-III sensors. Detailed thematic maps of soils, land use / land cover, hydro-geomorphology were prepared using the IRS IC/ID LISS-III and PAN data, supported with ground information. Following the above mentioned methodology critical areas for land treatment in the selected micro watersheds were identified.

After identifying the critical areas specific action plans were developed for land and water resource development in consultation with local farmers & cooperating centers. The land and water resource action plan of the Nipana Micro watershed, Akola Dt. Maharastra is shown in Fig.2.

Intermittent contour trenches

Recharged well with high water level

Conservation pit graded bund

Good crop of cotton beside recharged well

Cement gully plug

Cotton varietal trials

Fig.3 Implementation of action plans

Action items such as construction of water harvesting structures, soil conservation measures, crop improvement techniques, etc were identified and implementation of the plans was carried out in the selected micro watersheds (fig.3). The impact of the implementation was felt in all the micro watersheds by the farmers and there was an improvement in their socio economic conditions.

In Nipana micro watershed, Akola, new technologies like conservation pit graded bund (CPGB), Intermittent contour trenches (ICT) were developed.

Farm ponds – 5, Gully plug – 8, Cement Nala bund – 1, diversion ditches – 7, loose boulder structures – 2, were constructed and this helped in increasing the water levels in the dug wells. Three diversion ditches were prepared to remove excess runoff and to improve drainage network

. Adoption of improved practices like opening of alternate furrows and sowing across the slope, the yields of cotton varieties AKA-7, NHH-44 and AHH-468 have increased in shallow, medium and deep soils by 11.4 to 19.9% over farmers practices. Cotton varieties AKA-7, NHH-44 and AHH-468 sown in moderately deep soils have recorded the increase in the yield of 20.48, 31.20 and 48.05%, respectively over those sown on shallow soils. Gully stabilization on the bank of nallah was done by planting 300 bamboo seedlings.

CONCLUSIONS

Remote sensing and GIS have proven to be valuable tool in characterizing the watersheds in terms of their natural resources , creation of data base in timely and cost effective manner to arrive at optimum solution for various problems existing in the watersheds.

With the availability of high resolution PAN and multispectral remote sensing data, the scope of natural resources inventory of micro-watersheds has greatly improved at 1 : 12500 scale. The IKONOS and Quick Bird satellite data, should be evaluated for preparing various natural resources themes on 1 : 8000 scale (multispectral data) and 1 :4000 scale (PAN + multi spectral data).

The remote sensing technology could be used for monitoring the progress of implementation work and also to assess the impact of the implementation interms of increase in biomass, increase in cropping intensity, improvement in soil moisture due to soil conservation measures etc.

Participatory approach between the researcher and the farmers are envisaged to better understand the needs, problems and acceptance of the action plans. The technical assessment of the watersheds should address the soil and water conservation problems with respect to agricultural land, horticultural areas, wasteland, special problem areas (eroded areas) etc. and should stress upon food, fodder, fuel and fibre needs of the people in the watershed.

References

ClarkW.F. (1990) , North Carolina Estuaries- A pilot study for managing multiple use in the state's public trust waters- Study report, 90-10,Raleigh, North Carolina

FAO. (1990) Watershed Management Field manual: Watershed survey and planning, conservation Guide 13/6. Rome.

NRSA and Dept. of Agriculture Govt. of AP., (2001), Perspective planning of Yeliminedu Macro watershed, RR Dt. AP.

NRSA,(1996). Soil survey and land evaluation for agricultural land use planning in tribal areas of Andhra Pradesh.

Rao,B.R.M. and L.Venkataratnam . 1992. Utility of IRS-IA data in soil resource mapping - A chapter in natural resource management A new perspective. pub. by NNRMS. PP. 424-430

Rao,B.R.M., Sreenivas K and Fyzee M.A, and T. Ravisankar. 1998. Evaluation of IRS-IC PAN data for mapping soil resources. NNRMS bulletin (22) pp.68-71

Rao,B.R.M., T. Ravisankar Fyzee M.A and Dineshkar 2004 Prescription for problems of micro watersheds under crop production system using remote sensing and GIS techniques- Soils & Crops Published by NRSA pp316-327

Saxena RK., K.S.Verma,G.RChary, RajeevSrivastavaand AK.Barthwal. (2000) IRS-Icdata application in watershed characterization and management. Int. J Remote Sensing Vol.21 No.17, 3197- 3208

Tideman,E.M, (1999). Watershed Management- Guidelines for Indian conditions. Omega Scientific Publishers, New Delhi.

Decision Support Model for Watershed Development

R R Hermon

P Kesava Rao

Vasala Madhava Rao
National Institute of Rural Development
(Ministry of Rural Development, Government of India), Hyderabad, India. 500030.

ABSTRACT

The satellite pictures available in recent years have become more sophisticated; the analytical tools for processing these images, have also become more powerful. As a consequence, it is now possible to utilize satellite images in a such more meaningful and practical manner. One of the important application of Space Science is the utilization of satellite images for efficient Watershed development.

A very large number of Watershed projects have been approved and taken up in almost all parts of India – under programmes like DPAP, DDP and IWDP (of the Ministry of Rural Development) and NWDPRA of the Ministry of Agriculture. Projects have also been sanctioned under schemes of these Departments, State Governments and under bilateral and International Assistance Programmes. It is estimated that over 4 million hectares have been treated or are under treatment.

It would have been natural to presume that satellite imagery is now being directly utilized for Watershed plans. However, the strange and somewhat unfortunate irony is that the satellite images have not, at the field level, been directly utilized for facilitating the implementation of Watershed projects. The last mile is yet to be traversed.

By the use of Remote Sensing and GIS , with the existing analytical tools, Action Plans can be prepared which can be utilized directly by the field level implementing agencies watershed intervention programmes, which could have taken about a year's time to prepare the same by manual survey methods, reflecting the potential of use of such a RS & GIS methodology.

Retrospect

The satellite pictures available in recent years have become more sophisticated; the analytical tools for processing these images, have also become more powerful. As a consequence, it is now possible to utilize satellite images in a such more meaningful and practical manner.

Organizations like National Remote Sensing Agency(NRSA) in Hyderabad, India, Space Application Centers in the States and other professional Institutes have been putting together satellite images and preparing sophisticated maps with the potential of a wide range of applications. One important application is the utilization of satellite images for efficient Watershed development.

The Department of Land Resources in the Ministry of Rural Development, Government of India, has engaged NRSA for several projects for the preparation of satellite based maps with a view to facilitate watershed development. The National Informatic Centre(NIC) has also been provided funds for linking relevant data bases with the satellite maps to enable a more detailed analysis. A great deal of work has been done by these agencies

A very large number of Watershed projects have been approved and taken up in almost all parts of the country – under programmes like Drought Prone Area Development Programes(DPAP), Desert Development Programmes(DDP) and Integrated Wasteland Development Programmes(IWDP) (of the Ministry of Rural Development) and National Wasteland Development Programme in Rural Areas(NWDPRA) of the Ministry of Agriculture. Projects have also been sanctioned under schemes of these Departments, State Governments and under bilateral and International Assistance Programmes. It is estimated that over 4 million hectares have been treated or are under treatment.

It would have been natural to presume that satellite imagery is now being directly utilized for Watershed plans. However, the strange and somewhat unfortunate irony is that the satellite images have not, at the field level, been directly utilized for facilitating the implementation of Watershed projects. The last mile is yet to be traversed.

An attempt has been made at the National Institute of Rural Development to cover this last mile. We are happy to report that, using satellite images, with the existing analytical tools, Action Plans have been prepared which can be utilized directly by the field level implementing agencies watershed intervention programmes.

Process Methodology

To demonstrate that the satellite data can be utilized if it is properly analyzed and processed for watershed interventions, a validation exercise was taken up with reference to certain watersheds selected at random in Andhra Pradesh State of India. The only criteria were that

    (a)   Action Plans/Detailed Project Reports (DPRs for these watersheds) should have been prepared, and

    (b)   All the structures which are technically required for these Watersheds, have been identified (with reference to the contours, the soils and so on).

It was an added advantage that in several of these selected watersheds, the work also had been taken up, and in some cases even completed. A comparison of the structures identified as per the Action Plan/DPR and recommended through the satellite image-cum-data analysis process revealed a remarkable congruence. It was found that all the structures recommended through the satellite data analysis, were actually included in the Action Plan/DPR based on a field survey. In fact, the satellite analysis revealed an interesting aspect – certain structures were being taken up in the field even though they were not required.

This shows quite clearly that if the satellite images are properly processed and analysed, they can form the basis for taking up the physical watershed intervention activities in the identified area. This can be done much more quickly and will serve as a useful guide for the watershed teams and PIAs who are managing these activities.

It takes a team of two professionals about 15 days to complete the analysis of one watershed covering 500 hectares.

The processing and analysis of the satellite images is not an elementary or simple exercise – it requires sophisticated equipment and professional and experienced manpower. Data in terms of information, and of course the satellite images of the required magnification/scale are a pre-requisite. Simultaneously, certain information collected from the field is also required. The details of the equipment, the qualifications and experience of the professional manpower, the specifications of the satellite images necessary, the layers of information/ data used with these images and finally the information from the field, have been also estimated. The particulars of costs, resources, etc., have also been worked out. The experience has been that all of this can easily be accessible or acquired.

Operational Strategy

It is, therefore, feasible to take up such an exercise for all these areas whether satellite data (which is the starting point) has been compiled. From the Department of Land Resources alone, the NRSA has been supported to cover the priority states in the country. It is understood that the information for several parts of country is available and has only to be accessed to be utilized.

The Department of Land Resources has approved under IWDP, DPAP, DDP and recently under Hariyali, 662 Watershed Projects covering 5.08 million hectares.

The Flow Diagram illustrates the details of technical process of analyzing the data and the stages through which it passes; the process has been illustrated with diagrams and maps to amplify the important stages.

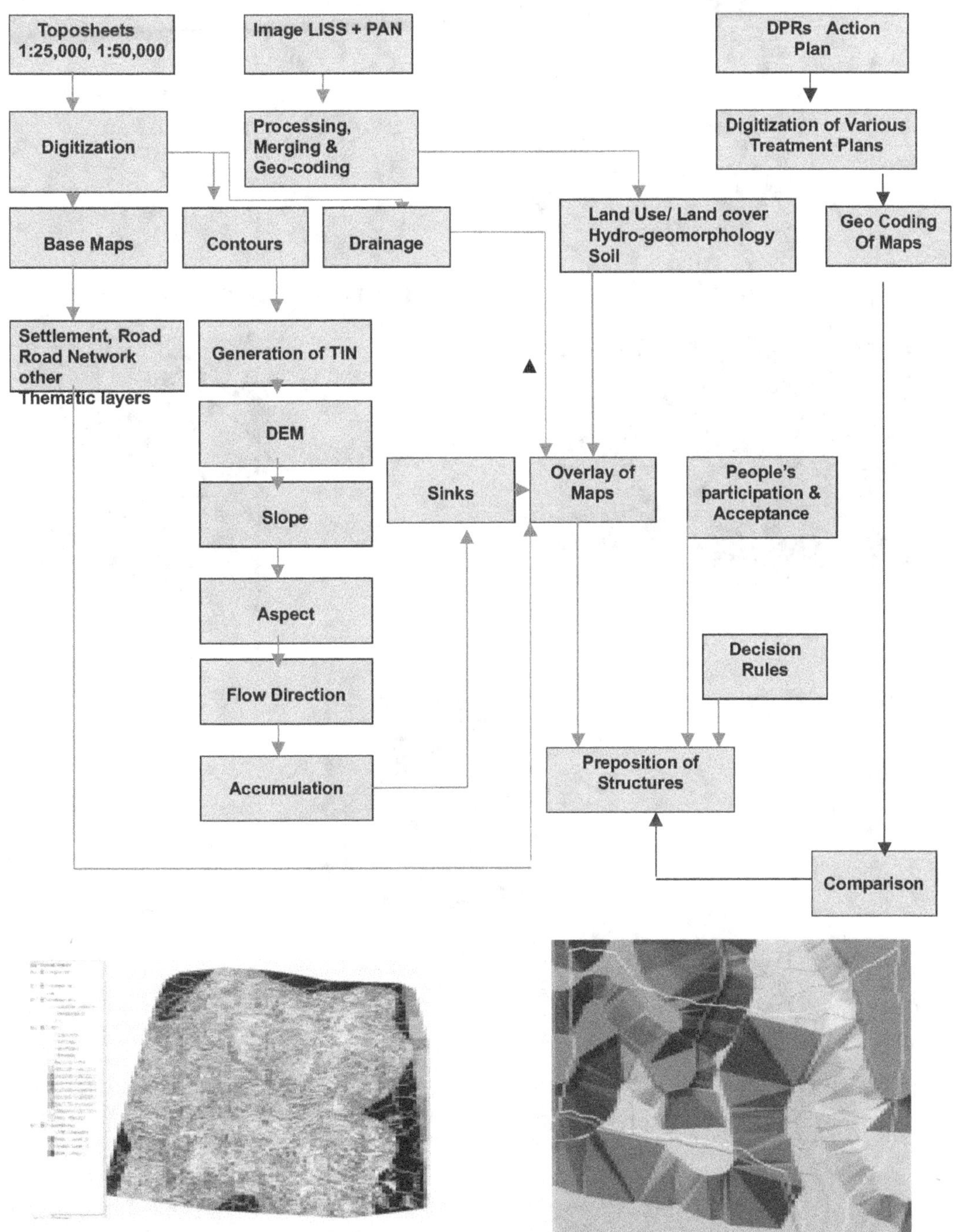

FLOW DIAGRAM

Satellite Imagery Draped Over Drainage

TIN with Drainage

167

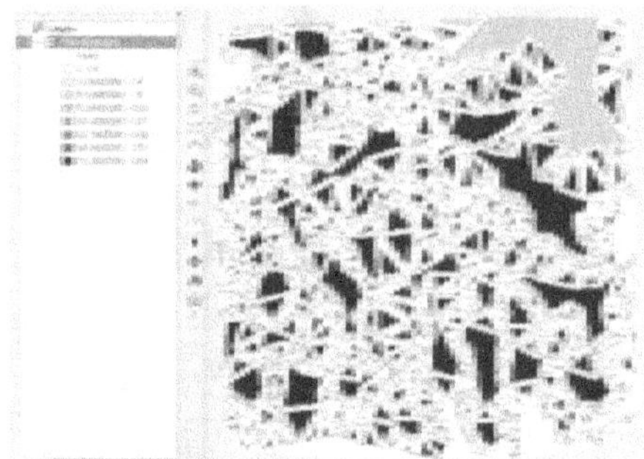

Flow Direction

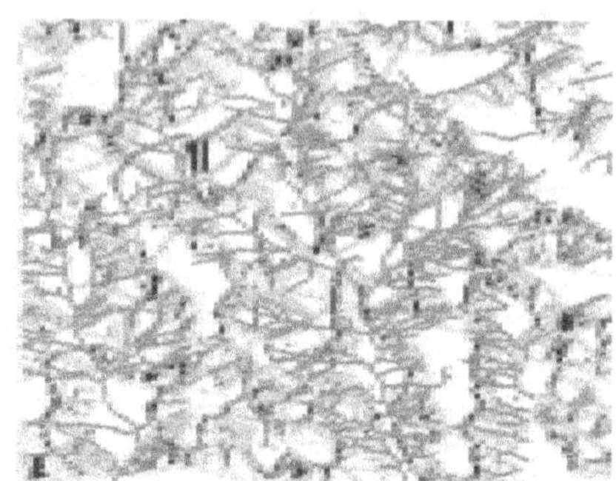

Drainage overlayed on slope aspect

Flow Accumulation

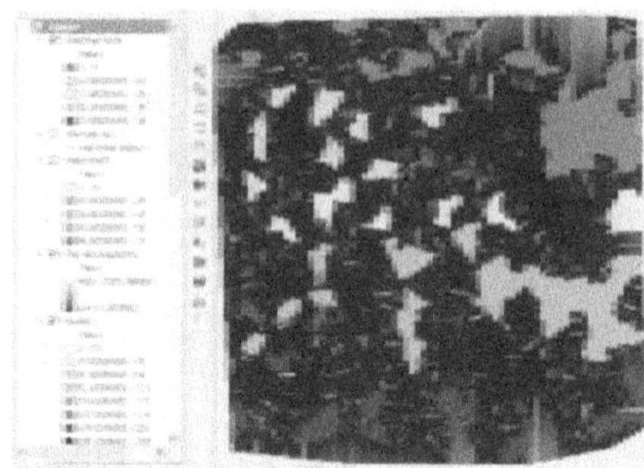

Sinks

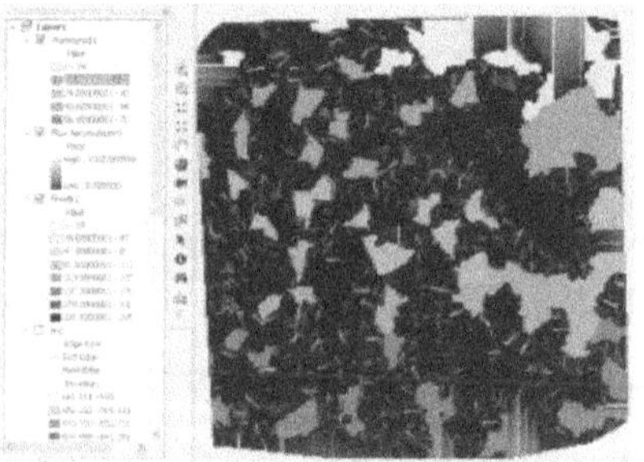

Area of Structures

CONCLUSION

This has been a pioneering exercise because it takes us through the last mile, without which the large investment in satellite imagery and data collection will remain redundant and limited to the theoretical.

It is for us to assess the potential of this opportunity and to take the fullest advantage. The present one day workshop is an attempt to demonstrate how the satellite images can be utilized for assisting watershed activities under the IWDP, DPAP, DDP projects.

Development Of Spatial Database In Digital Domain For Drainage Areas Upto Micro Watershed Level In Umnabad Taluk, Bidar District By Using Remote Sensing And GIS Techniques

K. Ashoka Reddy*, H. Honne Gowda, L. Rangaswamappa***, M. Viswanath****, and A.S. Rajashekar******

KSRSAC Dept. of IT & BT, SOI Campus, Bangalore

ABSTRACT

Bidar is one of the drought prone districts in Karnataka and frequently faces water scarcity. To overcome this situation both Central and State Governments have sponsored developmental programmes concentrating on watershed. Investments, time, money and monitoring the progress of developmental activities had become difficult since the area chosen for developmental activity was large. Hence concentrating on the developmental activity on a microwatershed basis has become inevitable. All India Soil and Land Use Survey (AIS & LUS) has delineated the entire country into various hydrological units starting with water resources regions up to watersheds, which are having a size range of one lakh ha.

In this study watersheds have been further systematically delineated and codified into sub, mini and microwatersheds, the size ranging between 500-1000 ha., using Indian Remote Sensing Satellite data IRS-IC/1D, LISS III and PAN merged digital data and Survey of India topo sheets of 1:50,000 scale.

Humnabad taluk spreads over two river basins viz., the Krishna and the Godavari. Presently there is an increased interest towards micro watershed approach in implementing the developmental programmes both by the Government and Non-Governmental organizations. A detailed spatial database is essentially needed for scientific planning towards conservation and management of water and land resources. The delineated sub, mini and micro watersheds provides spatial database for scientific planning of development programmes in Humnabad Taluk.

Introduction

Humnabad taluk Bidar district which is the northern most district of Karnataka State. It comprises of 87 villages. Rainfall in the taluk is low and earratic. There are no major perennial rivers flowing in the taluk. Over exploitation of ground water and poor industrial infrastructure is the existing scenario of the taluk. For planning and to provide a base for sustainable development in the drought stricken districts on long and short term basis, it is very much essential to generate database. Drainage map provides information on drainages, watershed boundaries and distribution of surface water bodies. Information on drainage such as length, width, order, and other informations are derived by drainage morphometry to understand and characterize hydrological response of a watershed. Drainage map has been derived from satellite imagery, Survey of India toposheets are used as reference. Location map of study area is given in Fig. 1.

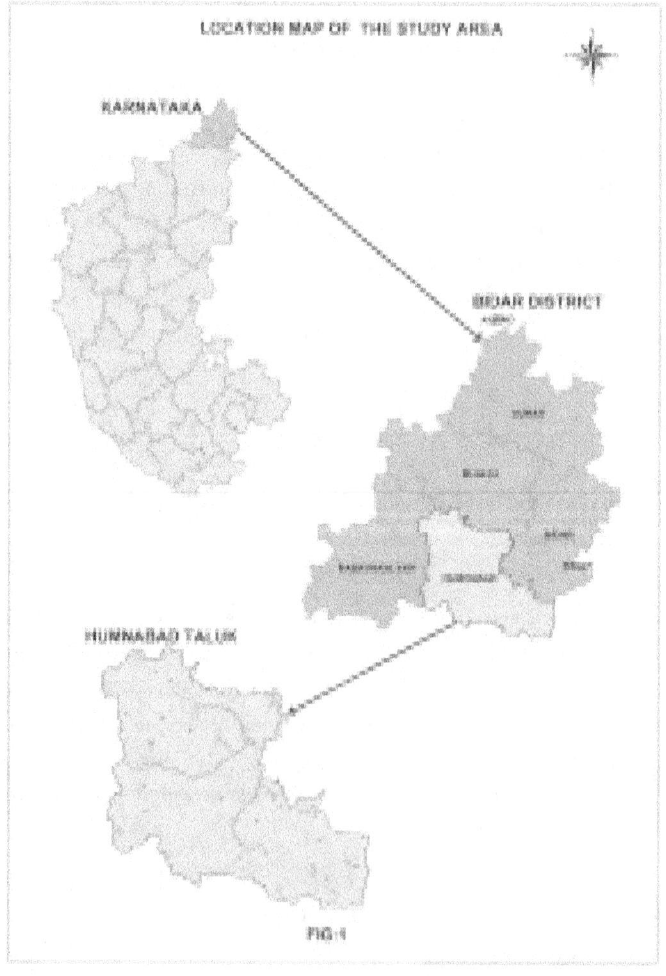

Objectives Of The Study

The delineation and codification of watershed into sub-, mini- and micro-watersheds has the following objectives:

1. To develop systematic delineation and unique codification as an extension of AIS&LUS (1990) and to provide a uniform base for the entire hydrological units of the taluk.

2. To develop a criteria for prioritization of micro, mini and sub-watersheds for implementation of watershed activity.

3. To develop a common base for various Departments/Agencies involved in developmental activities based on watershed approach.

4. To develop uniform basis for watershed characterization.

5. To develop Alpha-numeric codification which is simple and can be integrated with other layers in the digital domain.

Methodology

Methodology For Watershed Delineation

A framework of watershed is a prerequisite for giving practical shape to the systematic, scientific and rational approach for considering watersheds as units of planning and development. It is thus essential to have not only a hierarchical system of delineating bigger hydrological units into watershed, but also systematic codification so that each watershed could be identified as an individual entity without losing linkage with bigger units viz., sub-catchment, catchment, river basin and water resource region, to which it belongs.

Data Used

1. Satellite data (IRS-1C/1D PAN + LISS III merged image) on 1:50,000 scale

2. Survey of India toposheets are used as reference (1:50,000 scale)

3. The details from Watershed Atlas (AIS&LUS, 1990) of 1:1 million scale are transferred on to 1:250,000 scale base map.

4. Watershed Atlas of Karnataka 2005, KSRSAC, Bangalore.

Stages Of Delineation From Region To Watershed

All India Soil and Land Use Survey (AIS&LUS, 1990) has developed a 5-stage delineation system demarcating Water Resource Region, Basin, Catchment, Sub-catchment and Watershed. The size of watershed is determined by the size of stream or river of interest or the point of interception on the stream or river such as a dam, barrage, etc. Keeping in view the hurdles in managing large areas (watershed-wise), KSRSAC has further delineated the watersheds into sub-, mini- and micro-watersheds by adopting the methodology suggested by the Integrated Mission for Sustainable Development Programme of ISRO (NRSA 1993).

Stages Of Delineation From Watershed To Sub-, Mini- And Micro-watersheds

Based on the criterion – the ridge line method – the All India Soil and Land Use Survey has given 5-stage delineation (AIS&LUS, 1990). The manual of Integrated Mission for Sustainable Development (National Remote Sensing Agency, 1993) provides the guidelines for delineating sub-, mini- and micro-watersheds. In order to facilitate planning and development at micro-watershed level, in the present study 8-stage delineation has been carried out by further delineating the watersheds into sub-, mini- and micro-watersheds. The watershed map showing sub-, mini- and micro-watersheds is prepared on 1:50,000 scale maps. The 8-stage delineation is as follows:

Sub-catchment: The catchments are further divided into a number of sub-catchments which comprise mainly smaller tributaries and streams.

Watershed: Each sub-catchment has been divided into a number of watersheds.

Sub-watershed: Each watershed has been further divided into a number of sub-watersheds having an area in the range of 3000 – 5000 ha. Mini-watershed: Each sub-watershed has been further divided into a number of mini-watersheds ranging in size from 1000 to 3000 ha. Micro-watershed: Each mini-watershed has been further divided into a number of micro-watersheds ranging in size from 500 to 1000 ha. and are the smallest hydrological units.

The minimum area of watershed for implementation should be 500 ha. as recommended by Ministry of Rural Development and Employment and hence the delineation has been restricted upto microwatershed.

Delineation methodology is illustrated in Fig. – 2.

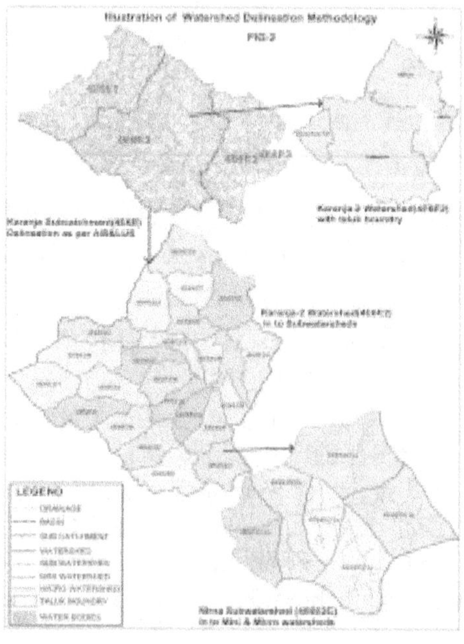

Codification Of Different Hydrological Units

Alpha-numeric symbolic codes consisting of a combination of alternating Arabic numerals and English capital alphabet have been used to designate various stages of delineation (up to Stage 7 i.e. mini-watershed). In the eighth stage of delineation (micro-watershed), English alphabet in lower case has been used.

WRRs are assigned Arabic numerals – 1, 2, 3, 4, 5, 6

Basins are assigned English alphabet – A, B, C, D…

Catchments are assigned Arabic numerals – 1, 2, 3, 4, 5, 6 …

Sub-catchments are assigned English alphabet – A, B, C, D…

Watersheds are assigned Arabic numerals – 1, 2, 3, 4, 5, 6 …

Sub-watersheds are assigned English alphabet – A, B, C, D…

Thus the sub-watershed will have codes such as 4D3D1A, 5A1A4C, etc.

Mini-watersheds are assigned Arabic numerals – 1, 2, 3, 4, 5, 6 …

Thus mini-watersheds will have code 4D3D1A1, 5A1A4C1, etc.

Micro-watersheds are assigned English alphabet – a, b, c, d…

Thus micro-watersheds will have code 4D3D1A1a, 5A1A4C1a, etc.

The coding of different stages (Stage 1 – 5) has been carried out from downstream upwards serially and from Stage 6 – 8, the delineation and codification has been carried out from ridge to valley .

Codification methodology is illustrated in Fig. - 3

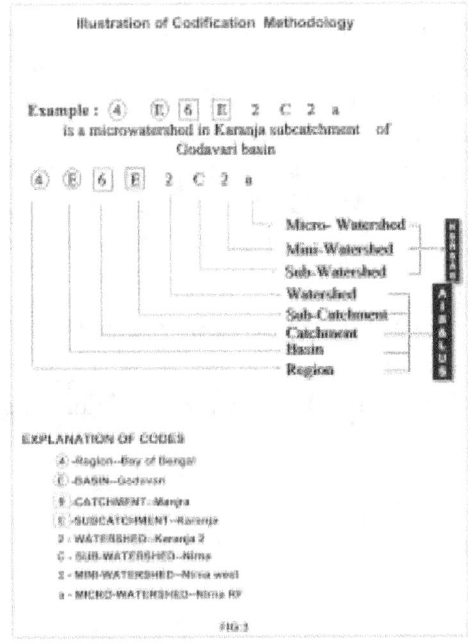

Preparation Of Watershed Map

The preparation of watershed map involves the following steps:

Boundaries of region, basin, catchment, sub-catchment and watershed from Watershed Atlas (AIS&LUS, 1990) of 1:1 million scale are transferred on 1:250,000 scale base map.

1. Extraction of drainage network from satellite image (IRS-1C/1D PAN +LISS III merged) on 1:50,000 scale and Survey of India toposheets are used as reference.

2. Boundaries from 1:250,000 scale base maps are transferred to the drainage map prepared using satellite data.

3. Delineation of watershed into sub-, mini- and micro-watershed based on hydrological unit and the extent of area on 1:50,000 scale.

4. In order to have unique code for the stages of delineation of watershed into sub-, mini- and micro-watershed, alpha-numeric system of codification is followed.

5. The sub-watersheds are named after the village present at the outlet or the name of the stream itself. The mini- and micro-watersheds are named after the village present within their boundaries. If more than one village is present, then the village that is at or nearer to the outlet is chosen for naming.

Results And Discussions

There are two river basins in Humnabad Taluk and they are Krishna and Godavari basins. The drainage area of Krishna basin in Humnabad taluk is 156 sq. km. This drainage area has been systematically delineated and codified into one Region, one basin, one catchment, one sub-catchment, one watershed, 8 sub-watersheds, 13 mini-watersheds & 41 micro-watersheds. The drainage area of Godavari basin in Humnabad taluk is 825 sq. km. This drainage area has been systematically delineated and codified in to one Region, one basin, one catchment, one sub-catchment, 3 watersheds, 25 sub-watersheds, 44 mini watersheds and 157 micro-watersheds.

During the development of spatial database it has revealed that 73 no. of micro-watersheds have been shared by adjacent taluks, which helps in the development of co-ordination between the taluk watershed development officers while implementing watershed development programmes.

The delineated microwatersheds of Humnabad taluk are shown in the Fig. – 4. The different Hydrological units are given in the Table – 1. The details of shared micro watersheds are given in table-2

MICRO WATERSHED MAP OF HUMNABAD TALUK
BHALKI
HUMNABAD
GULBARGA
CHINCHOLI
Legend
TALUK HEADQUARTER
METALLED
NATIONAL HIGHWAY
RAILWAY LINE
REGION
BASIN
CATCHMENT
SUB CATCHMENT
WATERSHED
SUB WATERSHED
MINI WATERSHED
MICRO WATERSHED
WATER BODIES
FIG.14

Table 1 Different Hydrological units of Humnabad taluk. Total

					Total
Water Resources Region	Bay of Bengal (4)				1
Basin	Krishna (4D)	Godawari (4E)			2
Catchment	Lower Bhima upto confluence with sina (4D5)	Manjra (4E6)			2
Sub-catchment	Bennithore – Mullamari on Left Bank Bhima (4D5B)	Karanja (4E6E)			2
Watershed	Mullaman (4D5B7)	Karanja 3 (4E6E3)	Karanja 2 (4E6E2)	Karanja 1 (4E6E1)	4
Sub-watershed	8	4	18	3	33
Mini watershed	13	4	38	4	57
Micro watershed	41	11	121	25	198

Conclusions

1. Remote sensing technology is very useful in updating the drainage network and also in the method for delineating and codification upto microwatershed.

2. Drainage map provides an effective basic material for planning conjunctive use of surface and ground water resources to optimum level.

3. The entire geographic area of 986 sq. km. of Humnabad Taluk. has been systematically delineated and codified into different levels of hydrological units (WRR to micro-watershed) the vital information about watersheds up to micro watershed has been summarized in the compendium of watersheds   which contains Basin-wise distribution of various hydrological units up to microwatershed in Humnabad taluk of Bidar District.

4. The study has resulted in creating spatial database. The database being in digital domain, the decision makers and users can access the data easily and develop scientific and systematic planning for the development of Humnabad taluk.

References

1. Watershed Atlas of India (1990), All India Soil and Landuse Survey, Dept of Agriculture and cooperation, Govt. of India.

2. Watershed Atlas of Karnataka (Sub, Mini and Micro watershed) 2005 KSRSAC, Dept. of IT& BT and S&T, GOK Bangalore.

3. PAN+LISS III Merged Satellite images, National Remote Sensing Agency, Department of Space, Govt. of India.

4. Toposheets of Survey India.

5. Technical Guide lines Integrated Mission for Sustainable Development (1993), NRSA, ISRO Dept. of Space GOI.

Irrigation System Performance of Nagarjunasagar Right Bank Canal

T. Saraswathi

Centre for Spatial Information Technology
JNTU, Hyderabad.

Introduction

Irrigation, land reclamation. Flood protection and use have an almost instantaneouseffect on environment and society. The cost of creating new irrigation infrastructure are rising rapidly and there are growing environmental concerns about large projects.

It is also essential to analyse and evaluate the performance analysis of the project devise tools for effective and to achieve better efficiency. In performance analysis there is a critically important distinction to be kept in mind between two complimentary points of view; Economic analysis and Financial analysis.

The economic analysis will help identified those project which makes the greatest contribution to over all economic growth. The economic analysis techniques are also neutral to income distribution and neutral to capital ownership as well. Although economic analysis will determine the amountof the income stream generated over and above the costs of labour amd other input it does not specify who actually receives it and in what proportion.

Financial analysis on the other hand, is concerned about income distribution and capital ownership. Financial analysis is important when we turn to a consideration of the incentive structure associated with a project investment. It will do us no good to have a project which is profitable from the standpoint of the whole economy if the individual farmers are unable to earn a living from their participation. Timing of returns, which the financial analysis will reveal is also important.

The techniques of financial and economic analysis of investment worth are only a tool of decision making. There are many non=quantitative and non-economic criteria for making project decisions. The usefulness of these techniques is to improve decision making process, not to substitute for judgement.

The economic analysis of a project focuses on the application of documented

Measures, cost – benefit ratio, net present worth and internal rate of return. Of these measures the latter two have their roots firmly in economics and do niot facilitate effective

"",

communication between experts from varios fields. On the other hand , the cost benefit ratio presents an easy picture which can be aborbe at a glance by an expert as well as a layman. This paper applied practical parameters to assess and evoluate irrigation system performance by computing cost – benefit ratio of water usage in part of Nagarjuna sagar right bank canal command areas.

Land use/land cover mapping

Land use/land cover categories and the pattern of their change is a prerequisite for planning, utilization and management of the land resouse sectors like agricultural planning, settlements and cadastral surveys, environmental studies and operational planning based on agroclimatic zones. Information on land use/land cover permits a better understanding of the land utilization aspects on cropping patterns, fallow lands, forests. Grazing lands, wastelands and surface water bodies, which is vital for development planning. The information requirements for land use planning comprise reliable, up-to-date and comprehensive data on physical, ecological and socio-economic resources. It is well established that remote sensing has the potential to make the most significant contributions in the area of land use data collection and more so in the agricultural land use.

Objectives

The major objective of the study is to performance of the Nagarjunasagar Right Bank Canal irrigation systems and monitoring.

The study area is restricted to Guntur disrict.

The following are the details of project formulations and objectives.

To apply remote sensing data for land use/land cover inventory in a part of Nagarjunasagar Right Bank canal command area in Guntur district of Andhra pradesh.

To delineate and classify areas under different land use/land cover categories as per classification proposed by NRSA using IRS Geo-coded FCC products on 1:50.000 scale.

To estimate the areas under different land use/land cover categories and to suggest appropriate recommendations.

To apply benefit-cost ratio assessment methodology and evaluate the irrigation system of Nagarjuna sagar right bank canal command area.

To evaluate Gross Benefit-Cost ratio of water usage in the Nagarjuna sagar right bank canal command area.

Along with the IRS data collateral data in the form of Survey of India toposheets has also been used. The present study involves preperation of land use/land cover maps of parts of Nagarjuna right bank canal command area.

Study area

Study area forms a part of the Nagarjuna sagar right bank canal command area. The study area is spread over in the extreme eastern parts of Guntur disrict of Andhra pradesh between 16degrees 45' North latitudes and 80 degrees 00' to 80 degrees 30' East longitudes. The Krishna river which flow from West to East forms the boundary.

Geomorphology the area represents simple features. The rock types exposed within the command area belong to Dharwar, Cuddapah and Lower Vindhyan systems.

The soils within the command area can be broadly classified as black soils and red soils. Above 62% of the study area is formed by black soil while 38% of the area is covered by red soils.

Methodology and mapping of land use/land cover using satellite imagery

The different phases of activity involved for interpreting, identification and mapping of land use/land cover using two seasons IRS false colour composite imagery on 1: 50,000 scale can be grouped as under:

Phase – 1	:	Selection and acquisition of data.
Phase – 2	:	Preliminaryvisual interpretation.
Phase – 3	:	Ground data collection and verification.
Phase – 4	:	Final interpretation and modification.
Phase – 5	:	Area calculation and estimation.
Phase – 6	:	Final cartographic map preperation and reproduction.

The above phase of activity can be executed by adopting the following steps:

Phase – 1 : selection and acquisition of data

1. IRS standard FCC imagery generated using a combination of spectral bands of 2,3,4 on a scale of 1:50,000 scale.

2. secondary data comprising of Survey of India topographic maps on 1:50.000 scale or on forest maps showing notified forest boundaries, other ancillary maps and stastical data are necessary.

Phase – 2 : preliminary visual interpretation

1. check that the study area to be interpreted is covered with respect to the available satellite imagery and the topographic maps.

2. spread the IRS FCC imagery on the desk table or light table and rest it firmly with weights.

3. for reference, indicate latitude and longitude from the scene 9if available, data and year of the scene, scene – row number, name of the district, etc.,)at the bottom corner of the overlay.

Phase – 3   :   ground data collection and verification

1.   The common area under crop during kharif and rabi seasons is the double cropped area. IRS imagery of both kharif and rabi seasons needs to be referred.

2.   If certain land use/land cover categories having similar responsse and spectral signatures appear side by side, then they become difficulty to be delineated separately. Such categories require detailed verification on ground.

3.   In case of inaccessibility and lack of information from other sources, the land use/land cover classes which look similar are generalized/merged with the next nearest class, which is homogenous and mappable. For example water logged and marshy land, salt affected and sands,jhum and forest blanks etc.,

Phase – 4   :   Final interpretation and modification

1.   After the collection on of ground truth, make the necessary corrections and modifications of land use/land cover boundaries are clarify the doubtful areas to the appropriate category.

2.   while modifying the thematic boundaries and transferring post=field details on to the pre-field interpreted map, use optical reflecting projector,if required.

3.   In the post-field corrected map all the details pertaining to kharif and rabi seasons crop land, double cropped area, follow land and other categories shown on a single base map.

Phase – 5   :   Area calculation and estimation

1.   Calculate area of different land use/land cover classes on thereproducible tracing film copy generated from the final master/original using millimeter polythene graph sheet and/or good grid planimeter for calculating areas.

2.   While computing the total or gross area under crop land, calculate and add the area under kharif, rabi and the area kharif+rabi (double) crop. To arrive at the net area sown, subtract from the total or gross cropped area the area under double crop i.e., area more than once. Also calculate simple percentage for the total cultivated area in the district.

Phase – 6   :   final cartographic map preparation and reproduction

1.   Prepare the fair drawing original (FDO) using mock-up (prepared to the specifications with legend, symbols and reference details) and the base map containing the final interpreted and transferred thematic details.

2.   Reproduce the FDO for multiple copies using ammonia paper. Use the ammonia copy for applying the appropriate colour. One coloured copy of the map may be used for area calculation.

Finally using GIS approach integration of thematic maps like slope, drainage and watershed maps are integrated with the land use/land cover layer.

Evaluation of Costs

Data analysis : The direct costs incurred on the water released into Nagarjuna Sagar Right bank canal can be broadly classified into two:

1. Cost of storage works (that is, cost of Nagarjuna sagar dam and reservoir).

2. Cost of distribution works (that is, cost of Nagarjuna Sagar Right bank canal structures).

Cost of storage works

Nagarjuna Sagar project is a multipurpose project catering to several purposes like irrigation, drinking water, power generation, recreation etc. this paper considers the analysis of water usage costs in the Right Bank canal for irrigation only.

Evaluation Of Benefits

DATA ANALYSIS: Rice is the principle crop cultivated in the command area irrigated as wet area. But the following diverse crops are cultivated in the command area considered as irrigated dry.

1. Cereals like Jowar, Maize, Bajra, Korra, Ragi, Vargu etc.
2. Pulses like Green gram, Black gram, Red gram, Bengal gram, Horse gram and Cow gram.
3. Edible oilseeds like Groundnut,sesame, Sunflower, Rape, Coconut, Mustard etc.
4. Non edible oil seeds like castor.
5. Cash crops like Cotton, Tobacco, and sugar cane.
6. Fibres like sunhemp and others.
7. Spices like coriander, Turmeric and Chillies.
8. Fruite including Citrusfruits.
9. vegetables.
10. flowers.
11. Mulberry trees.
12. fodder crops etc.

Analysis

Gross benefit-ratio of water usage in Nagrjuna Sagar Right bank canal command area.

$$\text{Benefit-cost ratio} = \frac{\text{Present worth of benefits}}{\text{Present worth of costs}}$$

Total discounted or present value of costs incurred on water usage in Nagarjuna Sagar Right bank canal = 4,321.30 Rs. In lacks.

Total discounted or present value of benefirs from command area of Nagarjuna Sagar Right bank canal = 32067.6 Rs in lacks.

$$\text{Therefore Benefit-Cost ratio} = \frac{32067.6}{4231.3}$$

$$= 7.421$$

Conclusions

Integrated analysis for resource management of a part of Nagarjunasagar Right Bank canal command area along the Guntur=Branch canal has been taken up for the present study using remote sensing techniques. The visual interpretation of sattelite data has provided a unique opportunity for studying land use and land cover and identifying agricultural land, forest areas, waste lands within the study area better tonal variation and pronounced contrast between various terrain features provided better identification of different land categories.

The land use/landcover map provides information about areas of established land use and also amount of land available for the development of new areas of agricultural, residential, transportation and other uses. Hence the information can be utilized for development plan and agricultural planning in the region.

Land use/land cover map can be applied to draw plans for better utilization of water for irrigation and pther uses in the region.

Agriculture being main occupation of the region, more land should be brought under cultivation through better management of irrigation water distribution with better irrigational facilities and scientific cropping pattern.

Land use/land cover map provides valuable information of forest areas and wastelands showing areas under degraded forest land and scrub land. Degrade forest lands must be developed by encouraging afforestation and thoroughly discouraging deforestation.

Wasteland areas of open land with or without scrub need immediate attention as these lands are prone to soil erosion. Efforts should be made to bring the scrub lands under cultivation and to convert lands affected by salinity/alkalinity into cultivable lands.

Biblography

1. Remote Sensing interpretation by Lillisand & Kiefer.

2. Principles of Remote Sensing by Sabins.

3. Remote Sensing and GIS for Environmental planning, edited by Dr.I.V.Muralikrishna.

4. Journal of Indian Society of Remote Sensing, volume 20,NO.4, December,1992.

5. proceedings of International Conference on Remote Sensing & GIS, 1992.

6. Cost Benefit Analysis by E.J.Mishan.

7. Economic Analysis of Agriculture projects by J.Price Gittinger

8. Water Resource planning and management by V.K.Sharma.

9. Irrigation Engineering by B.C.Punamia.

Catchment Capture : "A Geomorphic Analysis For Water Resource Appraisal" An Illustration From Mej River Basin (Rajasthan)

S. Padmaja

Professor, Department of Geography,
Osmania University, Hyderabad-7

ABSTRACT

Drainage channels are the most sensitive parameters to both intrinsic and extrinsic factors and are clearly reflected in the behavior of the channel form, drainage patterns and drainage extensions. This is mainly achieved through neotectonic movement which alters the drainage through river piracies. Mej River, a 5th order tributary of Chambal, located between 25^0-25^0 45'N and 75^0 30'-76^0 15'E, has under gone many geochronological events and exhibits a unique drainage anomaly. It can be termed as an antecedent river, with simultaneous drainage basin extension achieved through head ward sapping resulting in asymmetrical drainage basin.Mej was once considered a tributary of Banas. But later due to tectonic altercations has become a tributary of Chambal and is now in the process of a major river piracy where by the waters from Banas are likely to be diverted to Chambal through Mej. The shrinking and swelling nature of rivers and drainage have an impact on water resources. The wind gaps, misfit valleys and the adjoining catchments depict a picture of water resource depletion; where as the capturing stream and its extended drainage basin are likely to be associated with water resource enhancement. Under these changed conditions, the water resources need to be reappraised for better land use management. A few of these aspects are discussed for the Mej River basin.

Introduction

Drainage channels are the most sensitive parameters to both intrinsic and extrinsic factors, for the simple reason that the surface on which they flow can be subjected to upheavals and subsidence (neotectonics), altering the course of river. Further, the volume of water in the channels is a direct consequence of intensity of precipitation, leading to flood and drought conditions in different times. The change in the levels of the land due to tectonics and increase in the volume of discharge in the river will have a direct impact on the channel morphology and gradational activities of the river. This activity is usually more in evidence in the upper reaches of the stream segments, whereby the finger tip streams lengthen themselves by head ward sapping and erosion , encroaching on to the watershed of the adjoining basin and capturing the streams in the process. This catchment capture is a way of basin extension and basin elongation furthering the growth of drainage network and in the process altering the availability of water resources in the said region.

In the desert region of Rajasthan is seen such drainage, which is put to repeated changes. Palaeo drainage, intermittent and ephemeral drainage network are well evidenced in this part of the country. Banas and Chambal drainage basins too are the two pockets where such anomalous drainage forms are observed. Basinal asymmetry is the main character of these catchments and is seen through divide migration, drainage network adjustment and catchment capture.

Study Area

The Mej River is the main left bank tributary of Chambal, forming an oval shaped basin, over an area of 5500 kms, experiencing a sub-humid to semi-arid climate.

This basin is in a way transitional zone between North West lobe of Great Vindhyan basin and south east fringe of Aravallis. Geomorphically the basin extends from $25^0 0'$ - $25^0 45'$ North and $75^0 35'$-$76^0 15'$ east. North Western Vindhyan plateau is separated from South Eastern plain by Bundi Indergarh hills, which forms the back bone of the basin. Great Boundary Fault (GBF) movement was the root cause for thrusting Vindyans towards Gwaliors, thereby producing numerous faults in the basin. Folds are also common with Satur anticline standing out as an eroded anticline. The region is also characterized by numerous gullies which are developing on the Alluvium deposited by Mej and its tributaries. Geological structure of the area is complex resulting in intricate geomorphic forms.

In this repeatedly faulted terrain of the region, it is clear that the drainage becomes altered and modified to such an extent that, the study may no longer be an academic exercise but also will be a reappraisal of water resource availability and utilization with respect to existing land use conditions. With this backdrop in mind, the Mej river basin is taken as a unit of study with the intention of mapping the river capture of basin and to identify its impact.

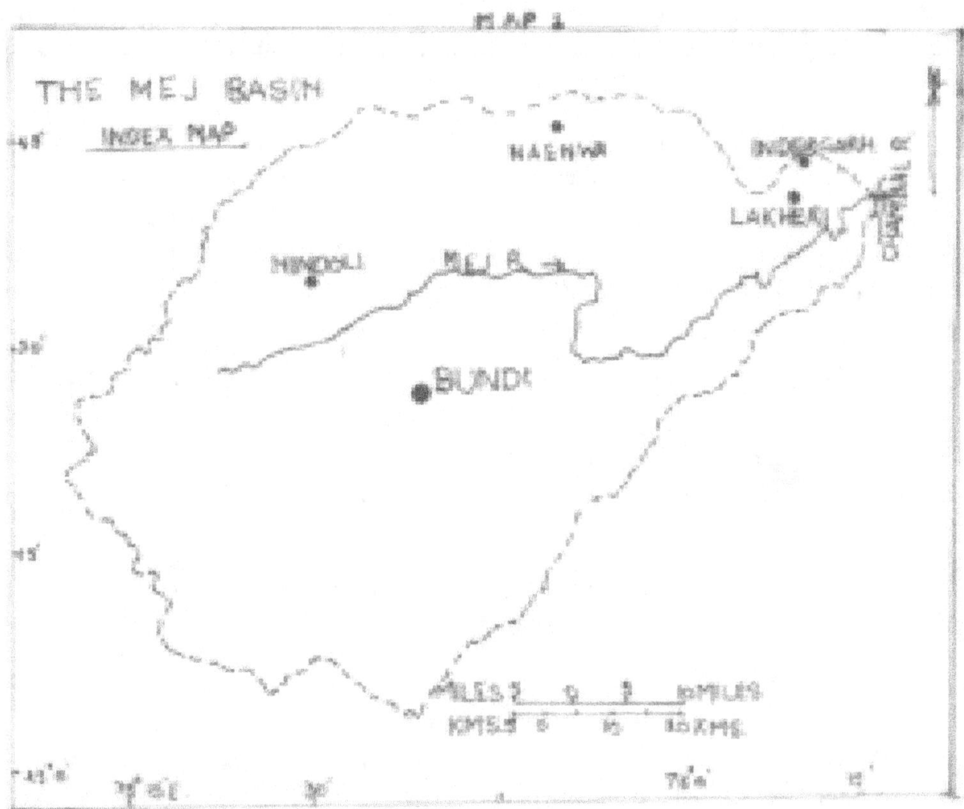

Data Base And Methodology

The broad database for the present study is :

    (a)   Toposheet (SOI) Nos of 45 O/SE, NE, 54C/NE, SW, 54 C/1,2, 45 O /10, 14.

    (b)   Satellite imageries of IRS, LISS, FCC and PAN

    (c)   Memories, Records, Reports, Census and Bureau of economics records.

    (d)   Field checks and observations.

Methodology adopted for this type of study is generation of thematic maps through remote sensing and GIS techniques.

Discussion

The Mej River is a 5th order tributary of Chambal river, having a fault trellis to dendritic, with an antecedent characteristic as it cuts into vindhyan range to meet Chambal. On the basis of geomorphic features, the Mej can be divided into three sections of Upper, Middle and lower, the total length of which is around 200 kms.

Upper Section : The River Mej originates from an elevation of 450 m to the north of Tikhi village. At the beginning, it flows in north east direction up to Dhanwara and then it takes an abrupt turn and adopts north western direction to almost north east till Gudha.

Upper course has a length of approximately 110kms and mostly flows over Gwaliors. In this section it is noticed that the Mej receives number of tributaries and it is also peculiar to note the number of tributaries are more on the left bank rather than on the right bank. The only single major tributary joining the Mej from right bank is Udayan Nadi, which alone carries water's of the Bundi hill. The major left bank tributaries of the Mej in the upper section are Balandi nadi, Goran Nala, and Machhli Nala.

Middle section : Downstream from Gudha, the river has cut across three successive Vindhyan ridges forming transverse valley. This is a tectonically deformed zone and the river is transversed by number of faults and accordingly forms into three transverse valleys. The total length of the course is around 20 kms. This shows the antecedent nature of the river.

Lower section : Downstream from Khatkar, the river enters into alluvial plain. Here the river exhibits an incised character and further develops ravines and gullies.

Valley cross profiles

Asymmetrical nature of the valleys is the highlight of the river. Left bank appears to be higher than the right bank, accompanied by under cutting. Sharma (1969) suggested that the Mej once might have been the tributary of Banas & was flowing parallel to the Bundi hills towards the north eastern direction. But its course later on is diverted by the Great Boundary Fault movement. This aspect clearly points out that the upper course of the Mej may be considered as consequent on the Gwaliors, middle course as transverse to the Vindhyan structure and lower course as strike valley. The geomorphic features of the Bundi hill clearly indicate that during Mesozoic period, the Mej was a tributary of Banas which up holds Sharma's (1969-73) explanation and it can further be suggested that it is a captured river by chambal. The formation of wind gaps in the Bundi hills is also due to capture of streams once flowing through them by the headwaters of Mej.

Drainage density (Dd) and Stream frequency (Sf) indicate higher values on the western bank rather than on the eastern bank, there by indicating higher dissection on the west than on east.

Resent observations on river piracy

In the upper reaches of Mej Basin adjoining to Banas basin, peculiar watershed behavior is observed. A bird's eye view of Banas basin depicts an extremely asymmetric basin shape due to uneven development of right and left bank tributaries of Banas catchment. A hypothetical tectono model by Sinha Roy reveals the possible stages of shifting Banas catchment.

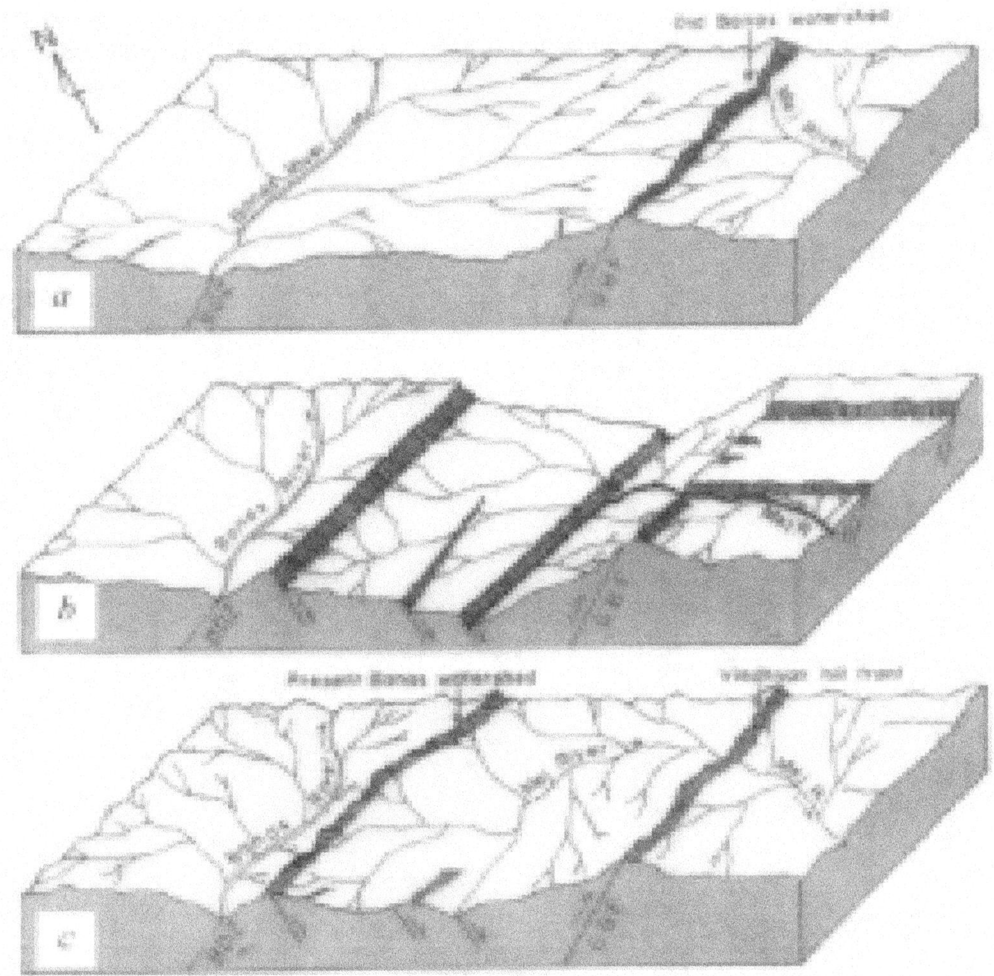

Source : Sinha Roy .S (2001) "Neotectonically Controlled Catchment Capture: An example from the Banas and Chambal drainage basins, Rajasthan"; Current science, vol 80, No-2, pp 293-298.

According to him "The Banas drainage basin in central Rajasthan is asymmetric in nature with respect to the Hortonian Banas river channel. The minimum distance of the channel from the watershed in the southeast segment of the Banas catchment, adjoining the Mej river sub basin of the Chambal catchment is only 7 kms. This feature is interpreted as having been caused by 25-30 kms migration of the Banas water shed from its initial position along the Vindhyan hills to the present position, due to superposition of the Mej river system and its head ward erosion in the Banas system in response to neotectonism .A part of the Banas catchment has been captured in this process by chambal river system. Since Mej River is very active, it is likely to pirate and behead a large segment of Banas River in future" (Sinha Roys-2001).

From the Basin map, it is clear that Banas catchment is highly asymmetric in nature in the sense that the main channel does not bisect the watershed. In its middle reaches, the average width of the basin on the left bank is 100 kms, while that on the right bank is 30 kms. Between Kachola and Devli the minimum width of the catchment is 7 kms near Jahazpur.

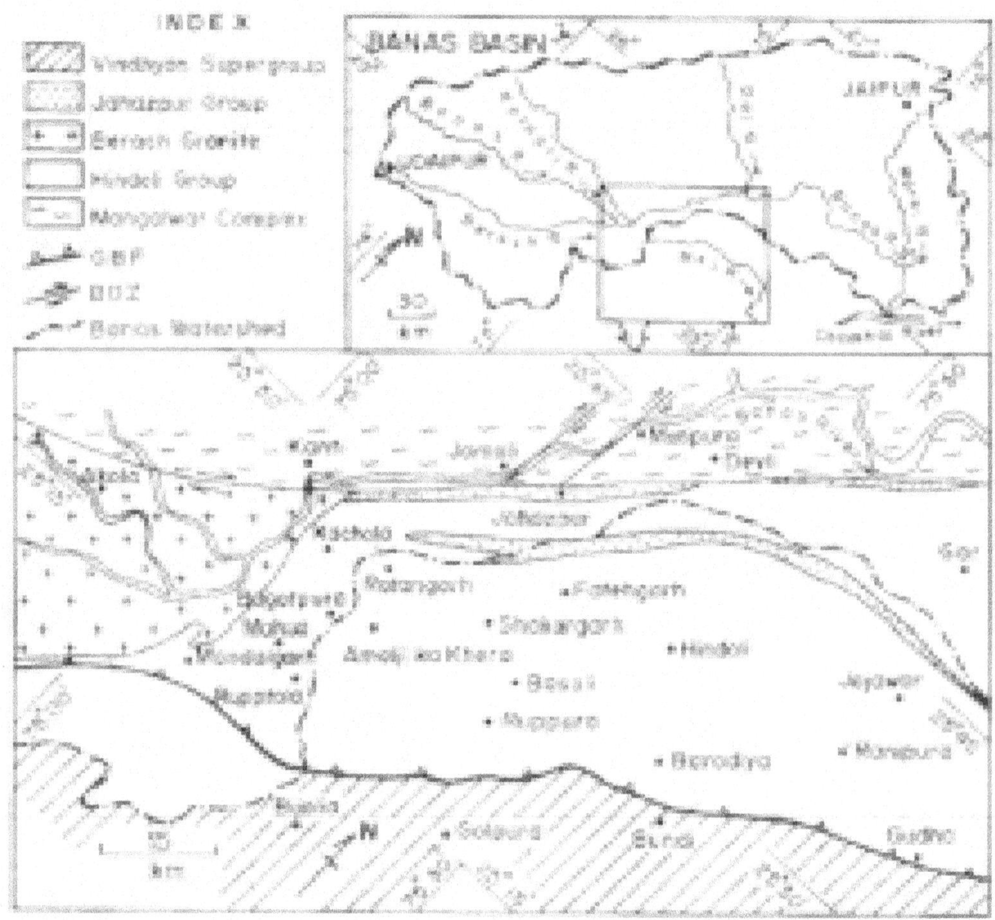

Source : *Sinha Roy .S (2001) "Neotectonically Controlled Catchment Capture: An example from the Banas and Chambal drainage basins, Rajasthan"; Current science, vol 80, No-2, pp 293-298.*

Evidences

A comparison of Drainage density and stream frequency values of the adjacent basins of Banas and Mej suggest that the values are higher for Mej than that of Banas. This also indicates a high competence and strong erosive power of the Mej River relative to the Banas. The Ground slopes also give an indication of steeper gradients for Mej, though it is a lower order river, compared to Banas, Dissected hill ranges and inselbergs are common depicting a part of old water shed.

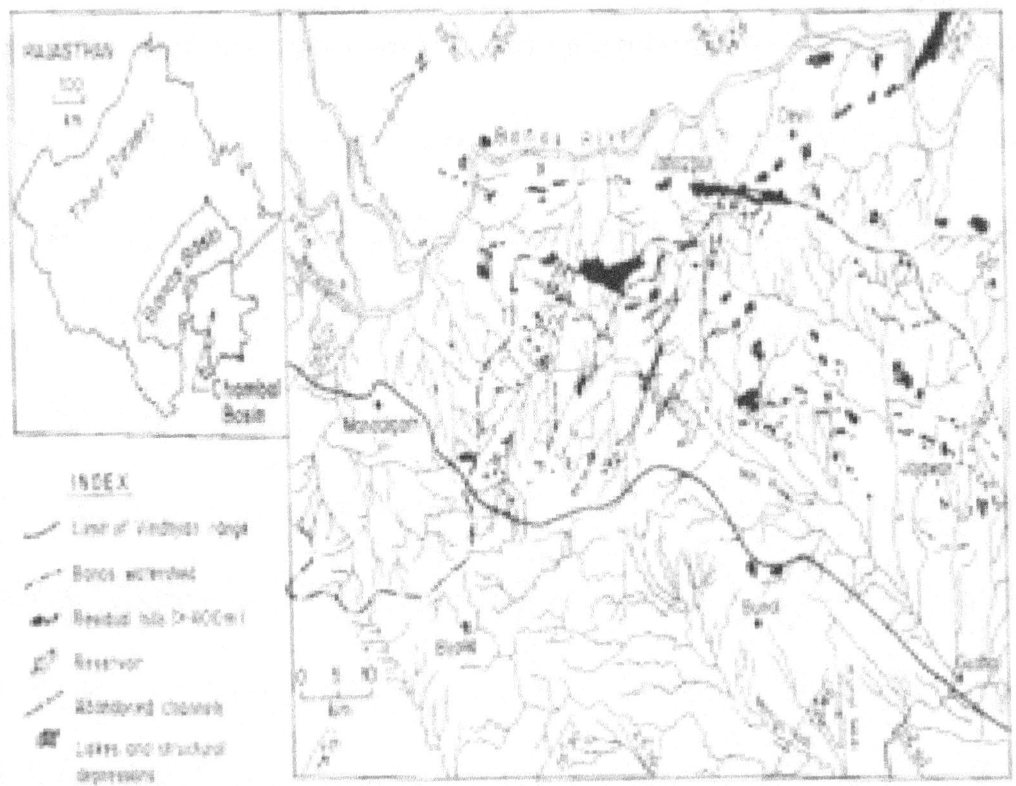

Source : Sinha Roy .S (2001) "Neotectonically Controlled Catchment Capture: An example from the Banas and Chambal drainage basins, Rajasthan"; Current science, vol 80, No-2, pp 293-298.

This region also has a number of sinuous palaeo channel, identified as relicts of Banas drainage, implying the shift of Banas River to North West which is subsequently taken over Mej.

Water resource appraisal

In the context of repeated drainage network adjustments, divide migration and catchment capture due to neotectonism, the Mej basin is put to repeated altercations. It is a river which was a tributary of Banas initially but captured by chambal due to tectonism. Similarly at present too, the river is heading for major capture of the Banas for which it was a tributary for sometime. The whole course of events is successfully changing the Mej Basin outline and with in chambal basins, inter basin alignment. The villages which seem to be effected negatively in terms of water resources are Devli, Motipura. The villages which are positively effected are – Fatehgarh, Shakargarh, Amalji Ka Khera, Ratangarh and region surrounding Hindoli. This is mainly predicted in view of grabenal floor, over which upper course of river flows. The elbow of capture may occur near Jehazpur, making the lower course waterless. However all this will take time and need to be assessed at length in detail.

References

1. Padmaja .S (1977) : "Geomorphology of Mej River basin", unpublished PhD thesis. (Rajasthan University)

2. Schumm, S.A. (1977). "The fluvial system"; John Winley and Sono, NY.

3. Sinha Roy .S (2001) "Neotectonically Controlled Catchment Capture: An example from the Banas and Chambal drainage basins, Rajasthan"; Current science, vol 80, No-2, pp 293-298.

An Integrated Approach On Watershed Planning And Management Through Artificial Recharge And Rainwater Harvesting – A Case Study

C Bhanu Prakash and Dinesh Kar

Speck SpatialTech Limited, Hyderabad, India

ABSTRACT

Groundwater is a precious resource with limited extent. Indiscriminate use of groundwater for various activities results in fast decline of the water resources. To prevent the aquifer from fast depletion, rainwater harvesting and artificial recharge of ground water is an established method. Integration of remotely sensed data and field survey data on a GIS platform provides convergent analysis of diverse data sets for decision making in groundwater management and planning. The advance technologies help in locating suitable sites for artificial recharge and rainwater harvesting. Artificial recharge technique utilizes subsurface geological formations for storage of substantial quantity of water received from surplus monsoon runoff. Other considerations for creating subsurface storage are favorable geological conditions and physiography including geomorphology which allows retention of substantial volume of water, in porous and permeable formations.

Speck SpatialTech Limited (SST) adopted a multi-disciplinary approach for developing a complete solution for planning watershed development through artificial recharge of ground water and rainwater harvesting in 2500 sq km area spanning across 3 districts in Chhatissgarh state of India. In total thirty two (32) watersheds of various sizes (ranging from 16 sq km to 248 sq km) falling in Mahanadi basin were delineated.

The study involved detailed surveys on natural resources, understanding hydrology of aquifers, evaluation of quality and quantity of ground water, analysis of socio-economic fabric, water demand assessment and its impact on ground water reserves. Resource inventories were generated as thematic layers using multi-spectral IRS P6 satellite image of LISS III sensor having spatial resolution of 23.8m. An integrated analysis on a GIS platform was carried out to identify sites suitable for artificial recharge and rainwater harvesting. Water resource development plan was generated for each identified site, describing type of recharge structure, detailed engineering design, cost benefit analysis and monitoring guidelines. Complete survey details, analysis and results were presented in the form of Watershed-wise Detailed Project Reports (DPRs).

Introduction

Water is an indispensable constituent of every day life and it is widely distributed in nature so that it may be available quickly and easily. However, with increasing population and rapid urbanization along with advent of modern technologies, water use has increased tremendously. Hence, there is a need for an early rational and practical policy for development of water resources, water use and its conservation i.e. optimal use of available water resources is essential for development of the country.

Remote sensing and GIS plays an important role in the study of natural resources and helps in planning water resources development. One of the greatest advantages of using remote sensing data for hydrological investigations and monitoring is its ability to generate information in spatial and temporal domain, which is very crucial for successful analysis, prediction and validation (Saraf, 1999). Remote sensing provides multi-spectral, multi-temporal and multi-sensor data of the earth's surface (Choudhury, 1999). However, the use of remote sensing technology involves large amount of spatial data management and requires an efficient system to handle such data. Thematic layers generated using remote sensing data like geology, geomorphology, land use/land cover, soils etc can be integrated in a Geographic Information System (GIS) framework and analysed with logical condition to generate ground water potential zone maps.

An integrated study covering the aspect of groundwater recharge is a crucial requirement of the present day (Choudhury, 1999). The present work is an attempt in this direction. Apart from the natural resources, geological and geo-physical surveys and hydro-chemical investigations, the study also takes an account of the social-economic fabric and water demand for the watershed area, and use remote sensing and GIS based analysis for identification of suitable sites for rainwater harvesting, artificial recharge. The study concludes with suggestion of suitable recharge structures with detailed engineering plans, maintenance and monitoring guidelines, cost estimates and cost-benefit analysis.

Objectives

The study was carried out to generate Detailed Project Report (DPR) on artificial recharge of groundwater and rainwater harvesting for Rampur watershed. The study includes;

- Identification, assessment and delineation of watershed boundary
- Assessment of the watershed for administrative set-up, physiographic features, drainage systems (streams and water bodies), slope, land-use/land cover, soil, geology/hydrogeology

- Water demand analysis

- Assessment of quality and quantity of water

- Identification of suitable sites for rainwater harvesting, artificial recharge structures

- Preparation of structure design, estimation and monitoring plan

- Generation of water resource development plan

Study Area

Physiographically Chhattisgarh state is divided into three distinct zones as Bastar zone, Chhattisgarh plains and Northern Hill region. The entire study area (32 watersheds) covers parts of the districts of Bilaspur, Korba and Korea, this falls in the Northern Hill region. The main river flowing in the state is Mahanadi and its tributaries are Seonath, Hasdeo, Mand, Arpa etc.

The Rampur watershed of Korba district falls in Mahanadi river basin. In this region the climate is predominantly sub-tropical with summer temperatures of around 29°C and winter temperatures of 21°C. The bulk of the precipitation is in July - September period [800 to over 1200 mm] with January - February precipitation of less than 50 mm. The river is one of the most active silt-depositing streams in the Indian subcontinent.

The district forms a part of central Indian Peninsular shield. Geologically the district comprises the rock formation of Archaean, Chattisgarh supergroup, Gondwana supergroup and Basaltic formations of Deccan traps.

Rampur watershed (Code 4G2C1r1) covers an area of 157 Sq km and is situated in the south-eastern corner of Korba district, Chhattisgarh state. The watershed is covered in Survey of India toposheet No.s 64 J/15 and 64 J/16 (1:50,000 Scale). The watershed lies between 82° 51' to 83° 00' longitudes and 22° 10' to 22° 20' latitudes.

Geology of the watershed

Geologically the area consists of Archaeans, upper proterozoic formations and Gondwanas. Archaeans are represented by granites and gneisses found in the SW corner of the watershed. South of granites, upper proterozoic formations are found. The upper proterozoic formations are represented by Chhattisgarh super group consisting of shaly sandstones, shaly limestones and limestones. The Gondwanas are represented by both lower and upper Gondwanas. The lower Gondwanas are represented by Talchir formations and Barakar formations, while the upper Gondwanas are represented by Kampti sandstones.

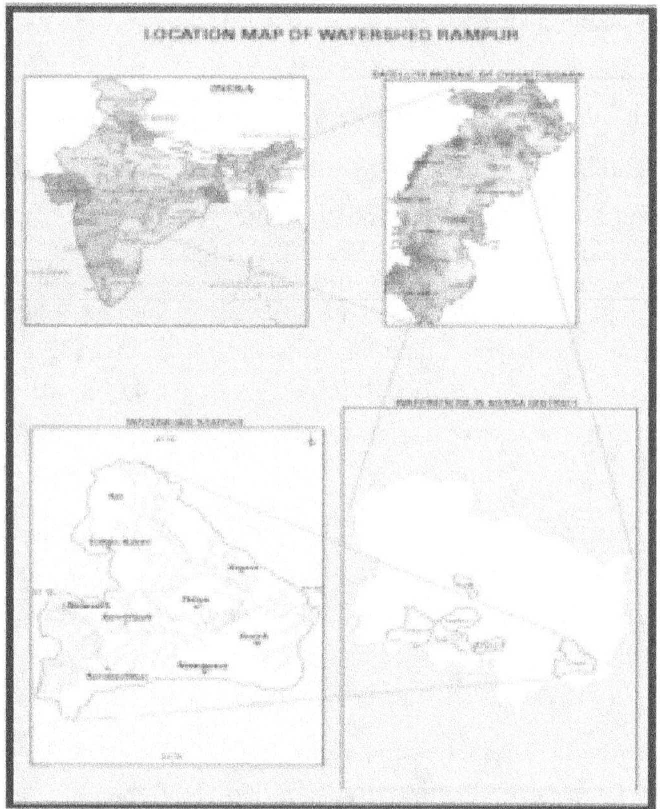

Fig. 1 : Location map

The Kampti bed belongs to Ranigunj series of upper Gondwanas. These Kampti beds comprise red and grey argillaceous sandstones and conglomerates with inter stratified shales. The beds contain patches and nodular of ferruginous material.

Data Used
Satellite data
LISS – III sensor data of Indian Remote Sensing Satellite IRS-P6 (RESOURCESAT) were acquired. Image data in multispectral bands as green (0.52 – 0.59 μm), red (0.62-0.68 μm), near infrared (0.77 – 0.86 μm) and having spatial resolution of 23.8 m were used. Two season images were procured from National Remote Sensing Agency Data Centre. Kharif season data acquired on 25th November, 2003 and Rabi season data of 29th February, 2004 were procured.

Ancillary data

A number of published maps and reports were used for the purpose of thematic layer generation as input. These are topographical maps at 1:50000 scale from Survey of India (SOI), geological maps at 1: 50,000/63,360 scale from Geological Survey of India (GSI), soil maps at 1:500,000 scale from National Bureau of soil survey (NBSS), maps from National geophysical research institute (NGRI) and reports from Public health engineering department (PHED), Central ground water board (CGWB), state ground water boards (SGB) were also used. Meteorological data like rainfall, temperature, number of rainy days etc. were collected from Indian Meteorological department (IMD).

Field observation data

Field data were collected for land use / land cover, socio-economic fabric and water demand. Demographic data related to the revenue villages of the watershed, Geophysical data for aquifer parameters and water level details were also collected through field surveys.

Methodology
Procurement of data

Satellite images for rabi and kharif season, existing geology maps, SOI topographical maps, soil maps, census reports and climatic details were procured from respective sources.

Delineation of watersheds

The drainage divides have been located by analyzing contour lines. From contour lines, attempt has been made to identify downside direction, and an array has been plotted to note down the direction. A divide line has been drawn where arrows show opposite direction. A stream is a line, where two arrows converge. This divide line has been drawn from the gauge location and the process continues until the whole watershed has been demarcated.

Preparation of base map

Base map for each watershed was generated using survey of India toposheet giving all required base information e.g. roads (metalled and un-metalled), cart tracks, railway lines, drainage, canals, ponds, important locations (Revenue Villages).

Geo-referencing and processing of data

Satellite imagery procured was rectified with reference to the SOI topographical maps. Image enhancement techniques were employed to make the satellite data more interpretable. Further the rectified image was subset based on the area of interest for each watershed.

Pre-field interpretation

Satellite data were interpreted for land use and land cover into different classes of agriculture, forestry, wastelands, built-up lands and water bodies. The minimum mappable unit at 1:50000 scale were kept as 3mm X 3mm. This layer is then exported into ArcInfo platform to check for dangle errors and label errors if any.

Ground truth collection

Field work for land use

Satellite imageries of both the seasons and the pre-field interpretation results were used for field verification. Location information was collected using GPS during the field visit. Tonal and textural differences were correlated with the terrain features. The units marked were modified according to the existing features on the ground.

Geo-physical survey

Geological/geophysical surveys were carried out in the field to understand aquifer geometry and groundwater status. Adequate number of geophysical soundings and pump tests were carried out for the watershed to generate maps for "depth to water table", groundwater contours, fence diagram etc. Available rock types in each watershed including structural and lithological details were studied as well.

Socio-economic survey

The study also focussed on the analysis of social profile, water quality, water supply, water consumption, awareness and participation in Water Harvesting Schemes. Such study necessitates the collection of the both primary and secondary data.

The primary data was generated through questionnaire surveys in all the villages. The questionnaire was structured so as to derive the information related to the objectives set forth. The questionnaire was divided into five modules through which information is extracted pertaining to

1. Village details
2. Population details
3. Water supply details
4. Water source details within and outside the village and
5. Awareness and Participation details related to Water Harvesting Schemes.

The secondary data is obtained through Census reports (1991 and 2001). Data pertaining to social profile is collected through Census reports, besides information is also gathered from the Sarpanch and district reports.

Based on the existing percapita and standard percapita consumption (@55 litres per person), water deficit or water surplus analysis was done for the year 2001. Based on the percapita deficit of water, four water supply priority zones are identified as given in the table below.

Table 1: Categorization of villages based on the existing water demand

S.No	Percapita Deficit / Surplus	Range
1.	Surplus	< 0
2.	Moderate	0 – 15
3.	Severe	15 – 30
4.	Acute	> 30

Water demand Analysis

Based on the 1991 - 2001 population growth rate, population for the years 2011 and 2021 was projected and water demand was forecasted for the same period taking per capita consumption at the existing rate for the rural areas.

Sample Collection & Analysis

Water quality analysis

Hydro-chemical analysis was undertaken by collecting representative water samples from each village in respective watersheds. A litre of water was collected from each village from available drinking water source with a special preference to groundwater source. The sample was subjected to analysis for standard physical and chemical parameters. The analysis helped in understanding the quality of drinking water in each watershed.

Soil sample analysis

Soil samples collected during the field visit were analysed for the infiltration rate and other chemical parameters.

Thematic maps generation

Physiography map & Slope maps were generated from the contours and spot heights taken from SOI topographical maps. Geological features like rock types were extracted from the existing GSI maps at 1:63,360 scale. The structures like lineaments were incorporated by extracting from the satellite images and SOI topographical maps. Geomorphologic maps were prepared using SOI toposheets, IRS P6 LISS III satellite imageries (FCC), and available literatures on geomorphology, geology of the area. Soil maps were generated from the existing maps from NBSS. The outputs were at 1:50000 scale. Based on the field observations the pre-field interpreted land use layer was modified and finalised. All drainages details were taken out from toposheets and further modified by satellite data. The units were checked for any errors of labelling. Groundwater prospects map was prepared using IRS LISS III FCC satellite imagery supplemented by ground data. Geomorphic, structural as well as lithological parameters of the watershed were taken into consideration. The groundwater prospects of the delineated geomorphic units were evaluated using available hydrologic characterise and aquifer parameters. Unit wise statistics for Land Use / Land Cover and Geomorphology were generated using Arc/Info platform and were reported.

Integrated Analysis and Site identification

Suitable Artificial Recharge sites were identified by an integrated analysis of various inputs from slope, geomorphology, soil, land use, geology of the terrain, aquifer characteristics and socio-economic study for water demand. Preference was given to the sites which were proximal to habitations with local precipitation.

The various types of structures suggested were Masonry/Boulder check dams and spreading ponds based on the site suitability and recharge potential.

Feasibility Study of the Proposed Site

Detailed engineering surveys were conducted to verify the proposed sites and type of structures. The gully section of the stream at the proposed site and catchment area for the sties were verified to study the feasibility the proposed site and structure.

Cost estimates

The cost estimates for various elements designed were prepared for different phases of construction and post-construction periods, which are essential for financial planning required during the construction work.

Cost benefit analysis

The recharge to ground water or effectiveness of water harvesting structures varies from 35 to 75%. The location of water harvesting structure in suitable sites and proper design plays an important role in effectiveness and efficiency of water harvesting structures. The evaporation losses are generally within 15% of the total storage.

Direct benefit due to any water harvesting structure may be attributed to additional recharge to ground water enabling additional area brought under cultivation, enhanced domestic and industrial water supplies.

METHODOLOGY

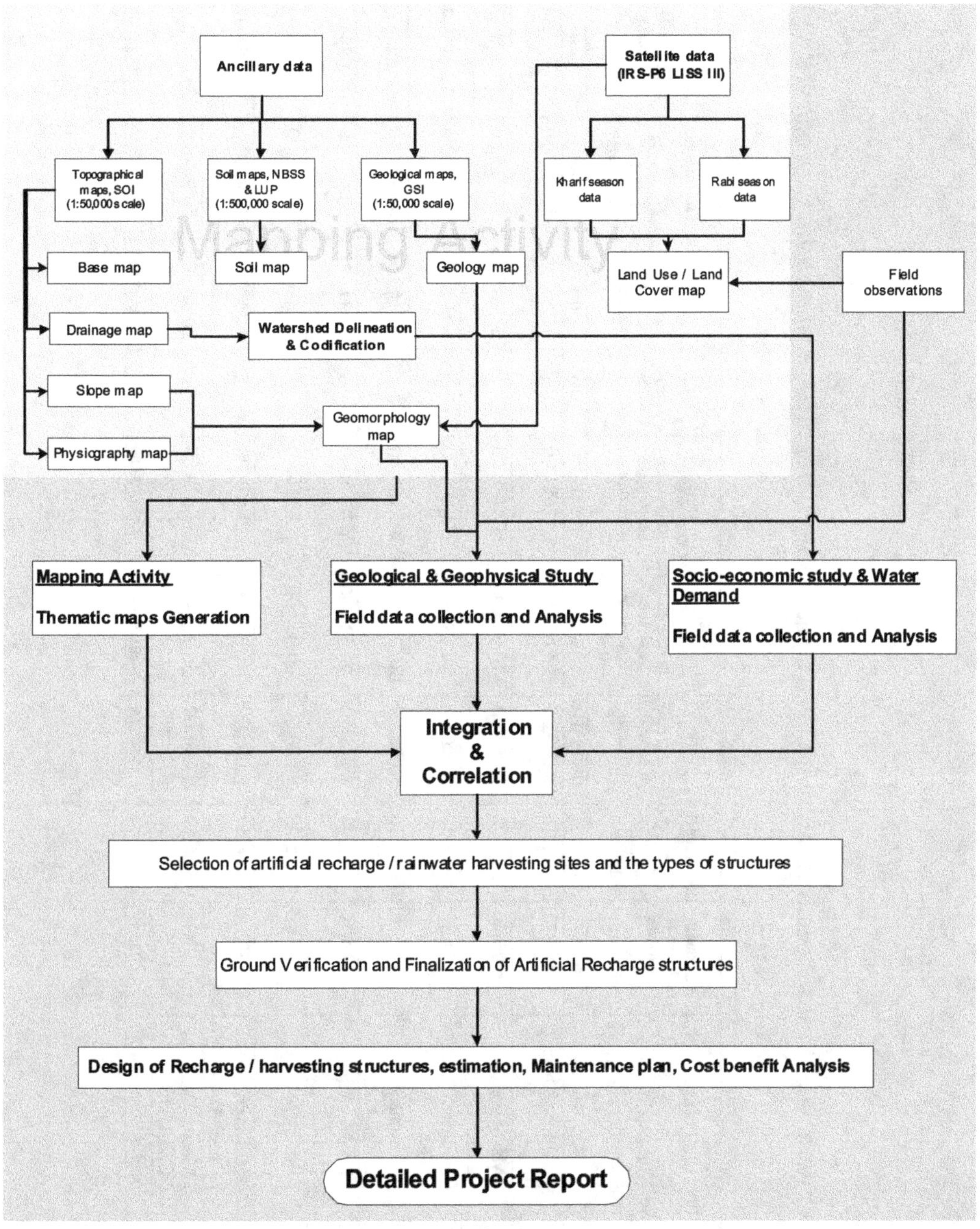

Table - 2 Cost - Benefit Analysis of Water Harvesting Structures Proposed in Watershed RAMPUR

Design No.	Type of Structure	Additional recharge to groundwater due to structure	Additional area brought under cultivation	Incremental income due to assured irrigation	Total incremental income due additional irrigation by structure	Constructional cost of water harvesting structure	Life of structure	Annual investment for construction	Incremental annual expenditure @ 10% of const. cost	Total annual investment	Cost benefit ratio
		TCM	(ha)	(Rs/ha/yr)	(Rs/Yr)	(Rs.)	(yrs)	(Rs.)	(Rs.)	(Rs.)	
1	Boulder Check Dam	58.77	11.75	8,000.00	94,032.00	537,681.00	25	21,507.24	53,768.10	75,275.34	1: 1.25
2	Masonary Check Dam	75.00	15.00	8,000.00	120,000.00	792,138.00	25	31,685.52	79,213.80	110,899.32	1: 1.08
3	Boulder Check Dam	21.30	4.26	8,000.00	34,080.00	274,401.00	25	10,976.04	27,440.10	38,416.14	1: 0.89
4	Masonary Check Dam	116.48	23.30	8,000.00	186,368.00	1,004,239.00	25	40,169.56	100,423.90	140,593.46	1: 1.33
5	Boulder Check Dam	14.71	2.94	8,000.00	23,536.00	207,779.00	25	8,311.16	20,777.90	29,089.06	1: 0.81
6	Boulder check Dam	25.02	5.00	8,000.00	40,032.00	271,250.00	25	10,850.00	27,125.00	37,975.00	1: 1.05
7	Masonary Check Dam	32.35	6.47	8,000.00	51,760.00	539,678.00	25	21,587.12	53,967.80	75,554.92	1: 0.69
8	Masonary Check Dam	32.36	6.47	8,000.00	51,776.00	509,672.00	25	20,386.88	50,967.20	71,354.08	1: 0.73
	Total/Average	375.99	75.20	8,000.00	601,584.00	4,136,838.00	25	165,473.52	413,683.80	579,157.32	1: 1.04

Conclusion

- Rampur watershed is triangle shaped and shows gentle slope towards southeast. The altitude varies from 340 m amsl at Junadih, located in the north, to 280 m amsl at Tenganmar located in the SE. A few disconnected hill ranges trending in NW-SE are also found in this watershed. Extreme slopes can be noticed in the northern and western boundary, a patch in the central portion of the watershed and also in the south west boundary of the watershed. This watershed is inundated by numerous streams and streamlets Chhindai Nadi is a major stream flowing broadly from W to E with a numerous streams and nalas from the all directions. Pansari nala is a major tributary of Chhinadai Nadi, which is located at the southern part of the watershed. The watershed is predominantly covered with well drained loamy soils in the entire region.

- As much as 80% of the watershed is covered by forest or wastelands. Almost half this area is under wastelands showing a vast stretch of dry land in the watershed. A small patch on the south western boundary comes under double cropping. The southern portion of the watershed is the main source of any cultivation with the area predominantly coming under Kharif cropping.

- Geomorphologically this watershed comprises Residual Hills (RH), Denudational Hill (DH), Pediment Shallow Dissected (PPS), Moderately Dissected Pediplain (PPM), Inselburg complex (PIC) and Pediment (P). The RH is found in NW, SW and S part of hill ranges. In these parts the RH is followed by DH. PIC, DH and RH are found in the hill ranges distributed on N, SW and S part of the watershed as described earlier. The plains show PPS and PPM. The river and stream courses give rise to moderately dissected weathering and thus give rise to PPM, while the uplands are represented by PPS.

- All the villages in the watershed conform to the Iron, fluoride and pH standards for drinking water. Turrikatra and Nawadih record excess CaCO3 and Nitrates respectively. Seven villages show excess dissolved solids in the ground water. Notably all these seven villages are found in the central part of the watershed inferring that the ground water in this region is more or less contaminated with dissolved solids. Hence it is proposed that more intensive study be carried out to confirm the groundwater quality in this region.

- Based on the 1991 - 2001 population growth rate, population for the years 2011 and 2021 is projected and water demand is calculated at the rate of 55 liters per capita. Regarding the projected water supply for the year 2011 and 2021 maximum demand comes from Bothi. Special mention is required to be made regarding Turrikatra and Tiladabra which are growing at a rate of 94% and 217%. There would be a heavy demand for increased infrastructure. Ghinara, at the rate it is declining may not exist beyond 2021, but this may not hold well if the population growth trend reverses. So, special attention should be provided to Ghinara to find out the reasons of such a huge decline. Discounting the demand from Ghinara, the minimum demand in 2011 and 2021 comes from Junadih, more so because of it relatively smaller size. A deeper analysis is required into the villages with a negative growth trend as a reversal in trend could shoot up the water supply demand and by all means one cannot plan taking into the consideration that the growth rate would remain negative forever.

- Type of structures suggested: The type of recharge structures suggested include Check dams, both masonry and boulder type. In total eight rainwater harvesting structures were proposed in the watershed.

- Cost benefit ratio : Assuming the life of a masonry / boulder check dams is 25 years, cost-benefit including annual incremental cost on construction and maintenance @ 10% has been worked out. As the direct cost of additional recharge in terms of rupees is not defined, benefit has been worked out in terms of additional income due to additional area brought under cultivation. Additional recharge from the water harvesting structure is worked out for 75% dependable rainfall and 15% of the catchment yield.

The details of the benefit by the additional recharge structures were furnished below:

- Additional recharge to groundwater due to feasible structures will be 375 TCM

- Additional area brought under cultivation is worked out approximately @ one ha. per 5 TCM and that works out to be an additional area of 75 ha.

- Incremental income due to assured irrigation is Rs. 8000/- per ha/yr.

- The cost benefit ratio for all the harvesting structure works out to be 1:1.04

References

Choudhury, P. R. 1999, integrated remote sensing and GIS techniques for groundwater studies in part of Betwa basin, Ph.D. Thesis (unpublished), Department of Earth Sciences, University of Roorkee, India.

Saraf, A. K. 1999, A report on Land use Modelling in GIS for Bankura District, Project sponsored by DST, NRDMS division, Govt. of India.

Jothiprakash.V, Marimuthu.G, Muralidharan.R and Senthil kumar. S.2003, Delineation of potential zones for Artificial recharge using GIS. Journal of Indian Society of Remote Sensing . Vol.31 (1); 37-47

Girish Gopinath and Seralathan.P.2004, Identification of Groundwater prospective Zones using IRS – 1D LISS III and pump test methods. Journal of Indian Society of Remote Sensing . Vol.32 (4); 329-342

Murthy K.S.R, Amminedu. E and Venkateswara Rao. V. 2003, Integration of thematic maps through GIS for identification of Groundwater potential zones. Journal of Indian Society of Remote Sensing . Vol.31 (3); 197-210.

A Wide Platform (www) for Implementing Applications of Geographic Information System

**Bhupendra Singh Purawat[1], Narpat Singh Rathore[2],
Gajendra Rathore[2] and Narender Verma[2]**
[1]Department of Information Technology,
ICFAI National College, ICFAI University,
[2]Department of Geography, C.S.S.H.,
Mohanlal Sukhadia University, Udaipur

ABSTRACT

Keywords : WWW, XML, Spatial Data Access, Digital Data Libraries, User Interface Design, Information Storage and Retrieval

The World Wide Web (WWW) has changed the way of accessing information in most organizations and gives a limitless space for sharing information without any geographical boundaries. A standardized set of protocols like TCP-IP (Transmission Control Protocol-Internet Protocol), HTTP (Hypertext Transfer Protocol), XML (extensible Markup Language) and other applications has allowed many organizations to share information among millions of users. Indian software industry has done well for providing services, solutions and software products; this is because of our large trained work force, our familiarity with the English language and the strong support of the government. The WWW and Web browsers have made the Internet user-friendly. The ability to integrate graphics, text, and sound into a single tool means that beginners do not have to struggle with such a complicated learning curve.

Though GIS is being widely used in the field of Geography; yet the basics need to be redefined and made simpler so that any person, even without technical background can understand it and derive the benefit of this technique. If this technology has to trickle down in our day-to-day activities it has to be the common man to appreciate and utilize these technologies. GIS, GPS and RS have wide applications in diverse fields like agriculture, archaeology, environment, geology, health, Business, information system, military, urban and rural development, transportation, telecommunication, power, water resource, natural and man made disasters, oil and natural gas, banking and insurance, mobile mapping etc. Internet plays a major role while implementing these techniques and provides a very good platform.

Due to lack of awareness at the level of layperson and problems of accessibility and usability of spatial data are significant and limited, therefore it increases number of users and application on limited resource that become bottlenecks. That's whey we have to develop easy to use and widely accessible spatial information system with user friendly and flexible interface that will be useful and effective particularly for inexperienced users. New users need to be aware of data services and be able to easily manipulate data. The paper is concerned with the design and development of user interfaces on WWW as a platform for improving the accessibility and usability of large and complex spatial data sets. This has led to academic and industry interest in developing technologies to efficiently and aesthetically render maps and map related data onto Internet browsers. But, there has been scant discussion and effort towards detailing all the components that are needed for a production-strength deployment of an Internet GIS system. WWW is increasingly used in geo-spatial research and spatial data services. This networked hypertext environment (over the Internet) provides an ideal platform for the GIS applications.

What is GIS?

Geographic Information System (GIS) is defined as an information system that is used to input, store, retrieve, manipulate, analyze and output geographically referenced data or geospatial data, in order to support decision making for planning and management of land use, natural resources, environment, transportation, urban facilities, and other administrative records. And links geographic information (where things are) with descriptive information (what things are). Unlike a flat paper map, where "what you see is what you get," a GIS can present many layers of different information.

To use a paper map, all you do is unfold it. Spread out before you, it is a representation of cities roads, mountains, rivers, railways, and political boundaries. The cities are represented by little dots or circles, the roads by black lines, the mountain peaks by tiny triangles, and the lakes by small blue areas similar to the real lakes. A digital map is not much more difficult to use than a paper map. As on the paper map, there are dots or points that represent features on the map such as cities, lines that represent features such as roads, and small areas that represent features such as lakes. All this information—where the point is located, how long the road is, and even how many square miles a lake occupies—is stored as layers in digital format as a pattern of ones and zeros in a computer.

Think of geographic data as layers of information underneath the computer screen. Each layer represents a particular theme or feature of the map. One theme could be made up of all the roads in an area. Another theme could represent all the lakes in the same area. Yet another could represent all the cities. These themes can be laid on top of one another, creating a stack of information about the same geographic area. Each layer can be turned off and on, as if you were peeling a layer off the stack or placing it back on. You control the amount of information about an area that you want to see, at any time, on any specific map.

Most computer technology is designed to increase a decision-maker's access to relevant data. GIS goes beyond mining data to give you the tools to interpret that data, allowing you to see relationships, patterns, or trends intuitively that are not possible to see with traditional charts, graphs, and spreadsheets.

More than that, a GIS facilitates designing models to test various hypotheses, derive results and test their efficacy with regard to the needs of various stakeholders. For example, a retail manager looking to build a new store can analyze consumer demographics and the locations of competitors in relation to potential locations in a spreadsheet view. GIS lets that manager visualize potential locations on a map along with drive-time analysis, environmental concerns.

Keeping these constraints and requirements in-mind while developing an Information System for geospatial data.

What Is Information System ?

An Information System is an organized set of components for collecting, transmitting, storing, and processing data in order to deliver information for action. We are living in the information society, so information is the key transforming resource in the new society that's not bounded by any geographic boundaries. The geographic data plays a major role while implementing any type of Information System for any nation. The GIS is utilized in almost all the disciplines, everywhere. The vast plethora of data and the digital maps generated by GIS are providing better analytical decision making tools that has made a considerable difference in planning today. Archaeology, Agriculture, Banking, Defence and Intelligence, power generation, Engineering, Surveying, administration, Fire/EMS/Disaster/Homeland-Security, Forestry, Health Services, Insurance, Education & research, Landscape Architecture, Libraries and Museums, Location Services, Marine, Media, Mining and Earth Sciences, Natural Resources, Real Estate, Retail Business, Telecommunications, Transportation, Water and Wastewater are some interesting areas where geographic data and GIS are being widely used.

The fundamental principle underlying the development of information systems is their use as a business tool, not merely something to have for the sake of using information technology. Thus, they should be developed on the basis of their ability to improve performance. However, such a rational means more than just profit and loss. Performance includes the benefits to end-users i.e. Government, employees, customers, and other people with whom the end-user interacts.

Relationship : Geographic Data Repository System And Information System

Linking location to information is a process that applies to many aspects of decision making in business and the community. Choosing a site, targeting a market segment, planning a distribution network, zoning a neighborhood, allocating resources, and responding to emergencies—all these problems involves questions of planning and decision making in specific geographic locations.

For anyone trying to evaluate information, the best way to do it is to view it on a map. Not just any map—intelligent digital maps made possible by geographic information system (GIS) technology. Everyone, including people who have never used maps to analyze data, is finding that maps make processing information much easier and more effective.

How Geographic Information System fits into an organization depends on the nature of the business and the reasons for its success. It is virtually impossible for Geographic Information System to be effective if it is developed independently of the objectives, values and goals of the business they support. The GIS should be integrated with other systems to help them, while taking decisions. Some decisions are a matter of routine when to take a certain action that is part of a fundamental operating procedure. Other decisions may call for determining what possible actions could be considered in particular situation. If the user is the prime concern, systems will most likely orient many applications to enhance user service. If cost control is the chief measure of an organization's success, the applications are more likely to be directed towards identifying and preserving cost advantages. If new products are the key to success, then systems should help identify or create them. Generally we might be surprised at how often situations arise that were taken into account during the design of a system and yet not effectively implemented. When this happens, and the information system can't accommodate the event, both the analyst and the system look bad, in a confrontation with a disappointed user. Thus Information System is a tool to support the system. Tools are only as good as those who use them. However, the proper use of tools can identify conditions and actions that might otherwise fall between the cracks. Using these tools doesn't mean that no surprises will occur. But their incidence will most likely be much lower. Geographic Information System can contribute the idea that lead to successful Information system.

Geographical Data is helping people make better decisions in many disciplines. It can be gathered and organized to support the generation of information inputs that are integrated in the business strategy of any organization. A geographic information system is not an end in itself. It is used to create useful information inputs that help organizations run better. It has saved hundreds of millions of dollars through increased productivity and efficiency. And it is just a beginning. GIS is helping thousands of organizations around the world. It is used on the Internet to organize its government for constituents. Simply touch a parcel on an online map, and the information for that location is available to you. *It can be accomplished by a perfect marriage between the modern vision of Geographic data repository System and Information System.*

Geographic Data Structure/ Spatial Data Structure

Geographic data structure is broadly defined as the location specific data pertaining to the physical, cultural and economic attributes of a region or regions arranged in a systematic and organized form either digitally or otherwise for use. It is necessary to describe the spatial location and distribution, as well as the attributes and characteristics, according to a specified form, termed a spatial representation model with a standardized data structure.

We use the term spatial data in a broad sense, covering multidimensional points, lines, rectangles, polygons, cubes, and other geometric objects. A spatial data object occupies a certain region of space, called its spatial extent, which is characterized by its location and boundary. Spatial data support in databases is important for efficiently storing, indexing, and querying of data based on spatial locations.

Geographic data are spatial in nature, but differ from design data in certain ways. Maps and satellite images are typical examples of geographic data. Maps may provide not only location information – about boundaries, rivers, and roads, for example – but also mash more detailed information associated with locations, such as elevation, soil type, land usage, and annual rainfall. Geographic features, such as state and large lakes are represented as complex polygons. Some features such as rivers may be represented either as complex curves or as complex polygons, depending on whether their width is relevant. Geographic information related to region, such as annual rainfall, can be represented as an array that is, in vector form.

Multi-tier Web-GIS Architecture

There is no one right way to develop a Geographic Information System, although there are ways to produce the right system for an application. Many variations on the development methods of GIS occur throughout the business community. Some methods are more successful than others, depending on when they are used, how they are applied, and who is involved in the development process. The ultimate determinant of success for a particular development method is the results obtained, not the theoretical "correctness" of the method.

It could be very helpful to apply database management or data warehouse techniques to store, retrieve and manage the data and models efficiently. Currently, most of the so-called GIS "Databases" are merely collections of data sets instead of being stored in and managed by a real Database Management System (DBMS). Most of these systems are stand-alone systems. Each application is running on one or several machines, and they are totally independent from each other. Thus it is difficult for different users to share and exchange information.

To achieve the system robustness, flexibility and resistance to potential change, the popular three-tier architecture is deployed in our system. The architecture is composed of three layers: the user interface layer, the application logic layer and the database layer. The three-tier architecture aims to solve a number of recurring design and development problems, hence to make the application development work more easily and efficiently. The interface layer in the three-tier architecture offers the user a friendly and convenient entry to communicate with the system while the application logic layer performs the controlling functionalities and manipulating the underlying logic connection of information flows; finally, the data modeling job is conducted by the database layer, which can store, index, manage and model information needed for this application. The three- tier architecture is used to increase performance, flexibility, maintainability, reusability, and scalability, while hiding the complexity of distributed processing from the user. These characteristics have made three layer architectures a popular choice for Internet applications and net-centric information systems.

Today's client/server applications resemble their ancestors so little that they have been given a new name, the *multi-tier application*, also known as *n-tier* architecture. In this model, processing is distributed between the client and the server, and business logic is captured in a middle tier. Most systems perform the following three main tasks, which correspond to three tiers, or layers, of the *n*-tier model. The three-tier architecture isolates each major piece of functionality, so that the presentation (user interface) is independent of the processing rules and business logic, which in turn is separate from the data. This model requires much more analysis and design up front, but greatly reduces maintenance costs and increases functional flexibility in the long run.

The third tier provides database management functionality and is dedicated to data and file services that can be optimized without using any proprietary database management system languages. The data management component ensures that the data is consistent throughout the distributed environment through the use of features such as data locking, consistency, and replication. It should be noted that connectivity between tiers can be dynamically changed depending upon the user's request for data and services.

Three tier architectures are used in commercial GIS distributed client/server environments in which shared resources, such as heterogeneous databases and processing rules, are required. The three tier architecture will support hundreds of users, making it more scalable than the two tier architecture.

Three tier architectures facilitate software development because each tier can be built and executed on a separate platform, thus making it easier to organize the implementation. Also, three tier architectures readily allow different tiers to be developed in different languages, such as a graphical user interface language or light internet clients (HTML, applets) for the top tier; C, C++, SmallTalk, Basic, Ada 83, or Ada 95 for the middle tier; and SQL for much of the database tier.

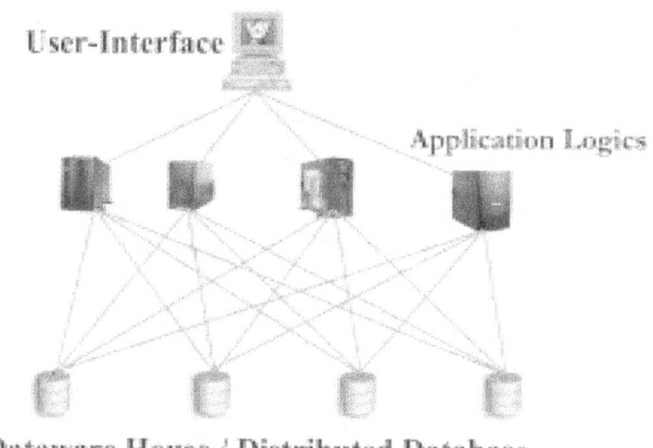

Fig. 5.1 Three-Tier Architecture

Top Layer: Interface (Desktop Applications, Internet Access Interface, Client)
The first tier is the user interface tier. This tier manages the input/output data and their display. With the intention of offering greater convenience to the user, the system is prototyped on the Internet. The users are allowed to access the system by using any existing web browser software. The user interface tier contains HTML components needed to collect incoming information and to display information received from the application logic tiers. The web visitors communicate with the web server via application protocols, such as HTTP and SSL, sending requests and receiving replies. In our system, the major web-scripting language exploited in designing the presentation layer is the Java Server Pages (JSP) or any scripting language. The intended system is prototyped into Internet; therefore, the design and implementation of the system user interface mainly becomes a job of designing and implementing web pages. The users can gain access to the system through any commonly used commercial browser such as Internet Explorer, Netscape, etc.

The end-user refers to people who are not professional information systems specialists but who use computers to perform their jobs. End-users are not all alike. Some may have never used a computer; others are intermittent users; still others may interact daily with GIS. Each end-user must be able to use the system easily and in a timely manner when required, even though its use may not be part of their daily routine. At the same time, the features of the system required to meet the needs of the infrequent user. The user Interface strives to balance systems features to suit the needs of all potential users.

Mid Layer: Application Logics (Logic as User requirement)

The middle tier encapsulates business logic. Some of this logic is application specific but a significant percentage is organization or even domain wide. Domain Engineering and Domain Analysis can be used to capture this inter-application commonality and create a set of assets that can be effectively reused in different application. The middle tier provides process management services (such as process development, process enactment, process monitoring, and process resourcing) that are shared by multiple applications.

The middle tier server (also referred to as the application server) improves performance, flexibility, maintainability, reusability, and scalability by centralizing process logic. Centralized process logic makes administration and change management easier by localizing system functionality so that changes must only be written once and placed on the middle tier server to be available throughout the systems. With other architectural designs, a change to a function (service) would need to be written into every application.

In addition, the middle process management tier controls transactions and asynchronous queuing to ensure reliable completion of transactions. The middle tier manages distributed database integrity by the two phase commit process. It provides access to resources based on names instead of locations, and thereby improves scalability and flexibility as system components are added or moved. Sometimes, the middle tier is divided in two or more unit with different functions, in these cases the architecture is often referred as multi layer. This is the case, for example, of some Internet applications. These applications typically have light clients written in HTML and application servers written in C++ or Java, the gap between these two layers is too big to link them together. Instead, there is an intermediate layer (web server) implemented in a scripting language. This layer receives requests from the Internet clients and generates html using the services provided by the business layer. This additional layer provides further isolation between the application layout and the application logic. It should be noted that recently, mainframes have been combined as servers in distributed architectures to provide massive storage and improve security.

Bottom Layer: (Data Mining, Data-ware House, Geospatial Data repository)

The database tier is responsible for modeling and storing information needed for the system and for optimizing the data access. Data needed by the application logic layer are retrieved from the database, and then the computation results produced by the application logic layer are stored back in the database. Since data are one of the most complex aspects of many existing information systems, it is essential in structuring the system. Both the facts and rules captured during data modeling and processing are important to ensure the data integrity. An Oracle database can be deployed in our system, and the Object Relational Model is applied to facilitate data reuse and standard adherence. Data analysis and modeling is a vital aspect of the database component. In our system, an object-relational design pattern is applied to model hurricane data. Object-relational model can assist the reuse of the database objects. An object-relational database schema is designed to facilitate the data reusability and manageability. The major advantage brought by the object relational concepts is the ability to incorporate higher levels of abstraction into our data models, while current relational databases are usually highly normalized models but with little abstraction.

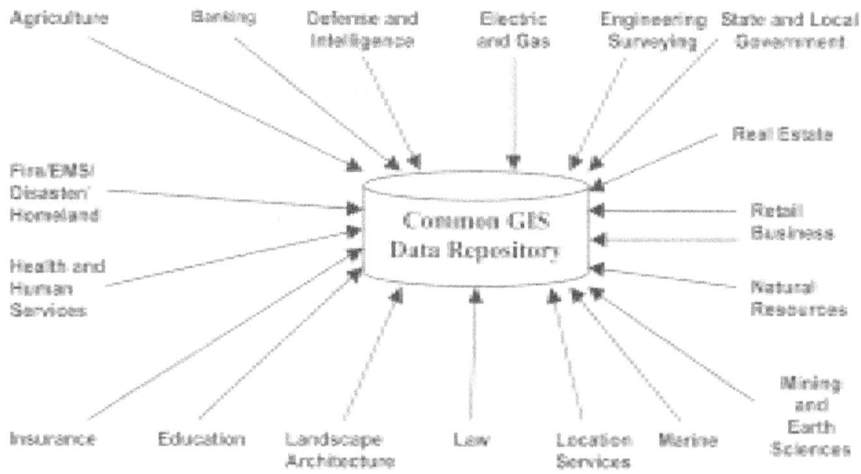

Fig. 5.2 Common GIS (Common Geospatial Data repository System)

Best Practice – Online Analytical Processing

Online Analytical Processing is a necessary part of spatial data mining, focusing on the end user's analytical supplies and computation process necessary to fulfill them. Just like, the popular operations of spatial OLAP are slicing and dicing, pivoting, roll-up and drill-down. We consider that OLAP should be integrated into Web-based GIS so that the capability of web-based GIS system will be improved greatly. The OLAP server supports multiple users, handling huge volume of data efficiently as well as supports well-to-do OLAP operations also. Multiple user access of spatial data repositories is very common now-a-days, especially to Web GIS applications. Typical OLAP operations are roll-up, drill-down, slice and dice, pivot. Spatial OLAP server should implement all of these functions. Generally OLAP runs against very large dataset. Hence efficient and effective access methods are critical. Furthermore, when it comes to spatial data, the situation becomes more complicated.

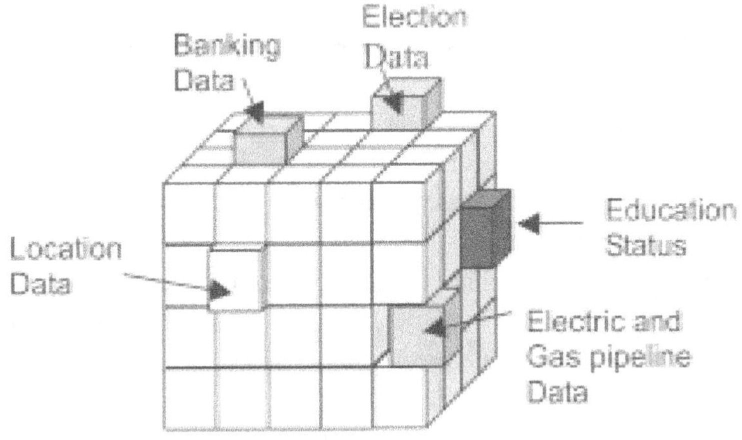

Fig. 5.3 OLAP (Multidimensional data)

Regarding data storage, the spatial data and its attribute are managed by geographical information systems in our system. The non-spatial data, metadata and concept hierarchy are stored in relational DBMS (RDBMS). Spatial OLAP server manages the materialized view which can shorten response time greatly. Data cube, which is constructed and managed by OLAP server, makes it possible for users to observe data from various concept levels. There are two major direction in implementing OLAP Servers, namely Relation OLAP (ROLAP) and Multidimensional OLAP (MOLAP). ROLAP extend traditional relational server to support multidimensional view while MOLAP utilize a direct way, such as, multidimensional array, to manage multidimensional information. ROLAP integrates naturally with existing technology and standards, which is reliable and scalable whereas MOLAP provide efficiency in storage and operations because of its direct representation of multidimensional data.

Conclusion

The most successful Geographic Information Systems, successful in terms of their benefit to businesses, originate with users. One reason is that the requests for these systems generally arise out of a business need that users perceive the need to solve a particular business problem, to handle routine business functions, or to monitor information as a way to prevent certain problems. The fact that business users often contribute the idea that lead to successful systems is as it should be, for the purpose of business, not to demonstrate the value of sophisticated technology. However, successful Geographic Information systems development is a joint effort. As significant as user contributions are, they draw out the best user ideas for discussion and analysis.

Geographic Information Systems is just another concept to analysis. The interplay among systems and subsystems and activities and functions in business is very real. Business strategy, government policies and Geography of state or country, these functions are not independent of one another in the actual day-to-day operations. Geographic changes influence Government policies; Government guidelines affect business policies. As in the proposed framework the system should be design to take an organization-wide view of operations and activities. Recognizing and understanding the relationships among the various functions of different departments/organizations and considering the impact of these relationships on the entire System throughout the development process will lead us to create the most useful Information Systems – useful because they fit the organization as it actually exists. Attention to the system approach to taking a global perspective.

References

- Abraham Silberschatz, Henry F. Korth & S. Sudarshan, "Database System Concepts", Fourth Edition, McGraw-Hill Higher Education, India, 2002.

- Andrew S. Tanenbaum, "Computer Networking", Pearson Education Publishing, India, 2003.

- Effy Oz, "Management Information System – Third Edition", Vikas Publishing House – India.

- Elias M. Awad, "Systems Analysis and Design", Galgotia Publications Pvt. Ltd, India, 2003.

- Gerald V. Post, "Database Management Systems – Designing & Building Business Applications", Tata McGraw-Hill Publishing Company-India, 2003.

- Gordon B. Davis & Margretha H. Olson, "Management Information Systems – Conceptual Foundations, Structure & Development", Tata McGraw-Hill Publishing Company-India, 2000.

- James A. O' Brien, "Introduction to Information Systems", Tata McGraw-Hill Publishing Company, India, 2004.

- James A. Senn, "Analysis and Design of Information Systems", Second Edition, McGraw-Hill Publishing Company, India,

- Jefferey A. Hoffer, Mary B. Prescott, Fred R. McFadden, "Modern Database Management – Sixth Edition", Pearson Education, 2004.

- Niklaus Wirth, "Algorithms + Data Structure = Programs", Prentice Hall of India.

- Raghu Ramakrishnan & Johannes Gehrke, "Database Management Systems", Third Edition, McGraw-Hill Publishing Company.

- Richard Fairley, "Software Engineering Concepts", Tata McGraw-Hill Publishing Company-India, 2004.

- Richard Y. Kain, "Advanced Computer Architecture – A system Design Approach", Prentice Hall India, 2003.

- Robert N. Anthony & Vijay Govindarajan, "Management Control System – Eleventh Edition", Tata McGraw-Hill Publishing Company-India.

- Roger S. Pressman, "Software Engineering – A Practitioner's Approach", Tata McGraw-Hill Publishing Company-India.

- Vladimir Zwass, "Foundations of Information Systems", Irwin McGrow-Hill Publishing Company, USA (Printed in Singapore).

Digital Terrain Modeling

R. Ramchandran
ADRIN, Secunderabad

ABSTRACT

Digital Terrain models play an important role in various applications. The current paper describes the definition, applications and methods of generation of DTM. Various techniques of visualization and accuracy assessment are also discussed.   A brief introduction is presented about the missions such as, SRTM, CARTOSAT-1, which are designed for obtaining the Digital Terrain Model.

Introduction

Terrain Models have always appealed to military personnel, planners, landscape architects, civil engineers, as well as other experts in various earth sciences. Originally, terrain models were physical models, made of rubber, plastic, clay, sand etc. Since the later 1950 the computer has been introduced into this area and the modeling of terrain surface has since then been carried out numerically or digitally leading to current discipline digital terrain modeling.

In representing the terrain surface, the digital terrain model (DTM) is one of the most important concepts. The word model usually means a representation and in many situations it is used to describe system in hand. It is reality scaled down and converted to a form, which we can comprehend. Model can be a conceptual, physical or mathematical model. The conceptual model is the model borne in a person's mind.

A physical model is usually an analog model. An example of this kind of model would be a terrain model made of plastic, rubber or clay. A stereo model of terrain based on optical or mechanical projection principles, which are widely used in photogrammetry, would also fall into this category. A physical model is usually smaller than the real object in geosciences.

A mathematical model represents a situation object or phenomenon in mathematical term. In other words, a mathematical model is a model whose components are mathematical concepts, such as constants, variables, function, equation, in equality etc. Mathematical models permit abstractions based on logical formation using a convenient language expressed in a shorthand notation, thus enabling one to better visualize the main elements of a problem while at the same time satisfying communication, decreasing ambiguity, and improving the chance of agreement on result. In addition, they help to generalize or apply the results of solving problems on the other areas. Mathematical model provides an opportunity to consider all the possibilities, to evaluate alternatives and to eliminate the impossible ones. They are tools for understanding the real world and discovering natural laws.

A good mathematical model should be accurate, robust, precise & general. In addition, it should be based on correct assumption & its conclusion should be useful, or inspiring & pointing the way to other good models.

Terrain can be defined as "a tract of country considered with regard to its natural features; or an extent of ground, region, territory etc.

'DTM is a more complex and all embracing concept not only involving height elevation but other geographical elements and natural features such as rivers, ridge line etc. in the narrower sense, a DTM only represents terrain relief.

There are several other terms e.g Digital Elevation Model (DEM), Digital Height Model (DHM) and Digital Ground Model (DGM) are frequently used to represent terrain, but they are different products.

'Ground' is defined as "the solid surface of the earth", a solid base or foundation, "a surface of earth". By contrast, height is given as "measurement from base to top" "Distance upwards" and "elevation above the ground or a recognized level" Elevation is defined as "angular height above the horizon; and height above a given level, especially that of the sea.

Thus a DGM is more or less has the meaning of a digital model of a solid surface". In contrast to the use of ground, the terms height & elevation emphasize the "measurement from a datum to the top" of an object.

If only elevation is the terrain information in the Digital Terrain Model – It can be called Digital Elevation Model. Obviously DEM is a subset of DTM.

In this paper, the emphasis is more on extracting & analyzing elevation information. The paper describes the need for DTM, methods of generating DTM from various sources. Sections follow these are on different ways of representing the terrain models, their accuracy & applications.

Why do we need DTM ?
DTM is fundamental model of the earth's surface, it has application in all earth relates sciences. DTMS are basic input for topographic mapping (contours), thematic mapping, orthoimage generation, map revision, image analysis etc. For civil engineering application

such as highway & railway design, cut & fill calculation & constructions of dams & reversion, precise DTMS are required.

In military engineering DTMs are draped with texture and other attributes to generate realistic scenery draped "fly through" are used for flight simulation. DTMS can also be used to guide cruise missiles. DTM is used to generate virtual battlefield. A number of parameters can be derived from the DTM for the battle field simulation, such as inter-visibility, shields of the landform, exposed distance of a moving unit, the doses shielding distance to the target and accessibility of the battle field.

DTM are useful for flood simulation, marine navigation agriculture management, communication network planning etc.

How to generate DTM?

The continents occupy about 150 million km^2 according for 29.2% for the earth's surface Relief varies from place to place, ranging from a few meters in flat areas to a few thousand meters in mountainous areas the highest peak of the earth is about 8884 m at Mount Everest, most oceans are kilometers deep while some trenches in the pacific are even deeper than 10,000 mts. The data sources for generating Digital Terrain Models are mainly:

1. Field surveying by using total station theodolite & GPS for direct measurement from terrain surface.
2. Photogrammetry by using stereo pairs of aerial or satellite images by analytical, analog or digital methods.
3. Radar-grammetry & SAR Inter-ferrometry
4. LIDAR (Airborne Laser Scanner)
5. Cartographic digitization by using existing topographic maps & digitizers.

Field Surveying

In traditional surveying techniques the position of a point is determined through the measurement of distance & angles. The traditional instruments are the theodolites and computerized total station. These techniques provide very accurate data but cost & speed of data acquisition is very high these techniques are applied to a small area only.

GPS are a popular technique for direct measurement of the earth's surface. They are replacing the traditional theodolites & total station

Photogrammetry

The basic principle of photogrammetry is to make use of a pair of stereo images to reconstruction the original shape of called stereo part 3 – D objects.

Stereo pair refers to two images of the same area imaged at two different perspectives with certain degree of overlap.

In aerial photography, there is generally a 60% overlap degree in flight direction and 30% between the flight strip.

Aerial photographs and images are cost effective & speedy techniques compared to ground surveys, aerial flights are affected by weather conditions & need support data as precise ground control points etc.

The quality of generated DTMs depends on various factors DTM can be viewed as physical models in stereoplotters. Measurements & further computations are done on these models to generate DTMs.

To process the photographic film on computers photographs are scanned by photogrammetric quality scanners and further processing is done in digital form now-a-days, digital images are acquired directly with airborne CCD matrix camera.

Aerial stereo pairs do not cover a large area. Aerial flights are affected by weather conditions and it is difficult to maintain a steady platform .

With aerial photogrammetry the area covered medium to large area can be covered.

In recent past very high resolution satellites with stereo capability are launched 1KONOS – 1 & IKONOS –2 Quick bird, Orbview & Indian satellite like CARTOSAT-1, which is fully dedicated for acquiring stereo images. Quickbird has spatial resolution of 61 cm in panchromatic mode & can acquire stereo in in-track direction. Orbview – 5. Scheduled to be launched in 2005 to acquire along track stereo at 41 cm ground resolution.

India's Cartosat-1 has spatial resolution of 2.5 m and has two cameras mounted at –5 & 26^0. The along – track stereo is formed with an objective to generate National DEM.

The space borne satellite stereo are cost -effective and cover large areas

LIDAR
The use of LIDAR for terrain elevation data has been established for sometime now. Operational systems are emerging. The usefulness of Airborne laser scanning system is that it is a direct measurement of elevation and is particularly found useful in complex tasks like acquisition of 3-D City models or the surveying & measuring the sag in power lines.

Cartographic Digitization
There are basically two cartographic digitization techniques, that is, vector based line following or raster based scanning. Digitization can be done either manually or by automated devices.

Radargrammetry and SAR Interferrometry

In practice, synthetic aperture radar, is widely used to acquire images. Images acquired by SAR are very sensitive to terrain variation. This is the basis for three types of techniques, that is, radargrammetry, interferrometry, and radar clinometry. Radargrammetry acquires DTM data through the measurement of parallax, while SAR interferrometry acquires DTM data through the determination of phase shifts between two echoes. Radarclinometry acquires DTM data through shape from shading. Radarclinometry makes use of a single image and the height information is not accurate for DTM. These techniques utilize the microwave region of electromagnetic spectrum. Thus data can be collected even in cloudy conditions or nighttime, **The Shuttle Radar Topography Mission** (SRTM) has collected the DEM of the 80 % earth' s landmass in just eleven days.. It is a joint project between the National imaging and Mapping Agency (NIMA) and the National Aeronautics and Space Administration (NASA). SRTM represents the first fixed baseline single pass spaceborne IFSAR (Interferometric Synthetic Aperture radar) technology with wide swath scanning SAR and dual frequency (C-band & X-band) coverage. The mission took place in Feb. 11-22 2000.

The mission objective was to obtain single pass interferometric SAR imagery to be used for DEM generation.

Coverage of earth's land surfaces provided between latitudes of 54 degree N and 80 degree S, representing 80 % of landmasses. The digital topographic map production objective (C-RADAR) calls for requirements which meet ITHD-2 (Interferometric Terrain Height Data –2) specifications, i. e. spatial pixel (30m X 30m) posting with 16 meters absolute linear vertical accuracy and 20 meters horizontal radial accuracy at 90% confidence level.

Virtually the entire planet was mapped to provide a topographic data over a period of roughly 255 hours during the STS-99 space flight. The SRTM-90 m resolution data has been released to the public. A second set of data (SRTM-30)- 30 m resolution exists and combines both C and X band of the sensing device, has restricted distribution.

SRTM data is a uniform matrix of elevation values indexed to specified points on ground. The horizontal datum is the World Geodetic System (WGS-84) and the vertical datum is mean sea level as determined by the WGS-84 Earth Gravitational Model (EGM-96) geoid. The elevation is w. r. to the reflective surface, which may be vegetation, man-made features or bare earth. The data is processed in one degree by one degree "cells". The edges of each cells are matched with the edges of adjacent cells to assure continuity.

The 90-m and 30-m resolution data sets have spikes, no-data regions etc, thus this data sets need to be processed before using in applications. A systematic database of SRTM data with suitable accessing method is needed. The options of using World Referencing Scheme-2 based and UTM zone based data sets should also be explored.

A comparison of various DTM acquisitions is given in Table-1.

Table-1 A Comparison of various DTM acquisition Methods

Acquisition Method	Accuracy of Data	Speed	Cost	Application Domain
Traditional Surveying	High (cm-m)	Very slow	Very high	Small areas
GPS survey	Relatively High (cm-m)	Slow	Relatively high	Small areas
Photogrammetry	Medium to high	Fast	Relatively low	Medium to Large area
InSAR	Low (m)(10 m)	Very fast	Low	Large areas
LIDAR	High (cm)	Fast	High	Medium to Large area
Map digitization	Relatively Low (m)	Slow	High	Any area size
Map Scanning	Relatively Low (m)	Fast	Low	Any area size

Representation of Digital Terrain Model

Digital terrain models are generated as irregularly set of point or regular grid of points. Through interpolation techniques irregular set of points can be interpolated to get regular grid of points Irregular set of points are stored as Triangular Irregular Network (TIN) Contour lines could be produced from either TIN based or regular DTMS. Contours can be considered as one way of representing DTMs.

Slope Shading and Hill Shading

Slope shading assigns a gray value to each pixel according to its slope value. Steeper the slope, darker the image.

In hill shading, a light sources is assumed, normally from the northwest. The facet facing the light is brightest & the facet facing away the darkest. Here, the idea is to portray the terrain variation with different brightness to illuminate so that shadow effects are produced, thus leading to stereo scopic sense.

Height based Coloring

Here, the term "height – based coloring" means to assign a color to each image pixel based on the heights of the DTM data. Two approaches are used – interval based and continuous coloring. The principle of hypermetric tinting (color layers) is to use different colors for areas with different altitudes. In continuous coloring, variation of gray tones represents variation of the terrain surface.

Rendering Techniques for 3 – Dimensional DTM Visualization

The basic idea of rendering is to produce vivid representation of 3 – D object. A surface is split into a finite number of polygons (or triangles in the case of TIN); all these polygons are projected onto the view plane of a given viewpoint; each visible pixel is assigned a gray value, which is computed based on an illumination model and the view point. In other words, rendering of DTM is to transform a DTM surface from a 3 – D to a 2 – D plane.

Texture Mapping For Virtual Landscape Generation

Mapping texture & other attributes on to the terrain surface creates a more vivid view of the terrain. Generally a 2 – D image array of the area is mapped on to the terrain model, further, annotations and contours etc. can also be added.

Animation Techniques for DTM Visualization

Animation is dynamic techniques for DTM visualization. The fundamental animation is the page flipping techniques , resulting in moves. First a number of frames of pictures are made and stored in computer memory then they are displayed on screen in sequences.

In terrain visualization, 'fly through" and 'walk through" are commonly used. The animated image sequence is produced in order of space, that is, by moving the view point along a certain tract.

Fly-through provides a continuous bird's eye view to the landscape. That is, the viewpoint is far above the terrain surface. Therefore the view point can be moved in any direction in the 3 – D space. Walk through mimics the human view while walking. Walking through can be considered as a special case of fly-through, that is the view point is low and its movement in vertical direction is restricted. The change in view point for fly – through or walk through can be controlled in various ways, such as using a mouse, keyboard, fixed route or freedom to roams.

With the advances in computer graphics capabilities of PCs. A seamless pan – view of DTM over a large area can also be seen on a desktop.

APPLICATIONS OF DTM

Applications in Remote Sensing and Mapping

DTMs have many applications in remote sensing and mapping, such as, topographic mapping (contours), thematic mapping, Ortho image generation and image analysis, map revision and so on.

Applications in Military Engineering

Flight simulation is a cost effective, safe and simple tool for pilot training. In addition to pilot training, flight simulation can also be used for mission planning and rehearsal.

In simulation, the DTM plays an important role, 3-D rendering techniques are employed to simulate the terrain. Texture & other attributes can also be mapped on to the DTM surface to generate realistic scenery.

DTMs can also be used to guide cruise missiles. This is done by matching the DTM surface stored in the computer with the real world sensed by detectors onboard the cruise missile.

Virtual Battlefield

The virtual battlefield is a simulation of a potential battlefield generated in computers which allows people to be involved. Battlefield simulation provide a dynamic and stereo environment, which can be used to recapitulate the battle, evaluate the results and gain experience. DTM is used to simulate the battle environment.

Flood Simulation

The flat areas of river basins are often flooded after heavy rain. Therefore, it is necessary to study flood risks. To do so, potential flood levels and velocity are the two major parameters to be considered. The DTM has been used to simulate floods. In such a simulation, with a given rainfall, the amount of water from different catchments can be estimated. After considering the capacity of the river, the amount of water to be accumulated can be computed. Then, the area to be flood can also be estimated.

Marine Navigation

The topographic surface is usually observed above sea level, but clearly the sub-sea terrain surface is important in various applications. Terrain model construction is often more difficult, as the overall surface form is often not available at the time of sampling. Observations of the sea floor are often made along ships tracks, giving a highly anisotropic distribution that requires special interpolation techniques to reconstruct plausible surfaces.

The most important marine application of DTM is in ship navigation. The dynamic intersection of the tidal sea surface is used for a collision-avoidance system.

Applications in Civil Engineering

The development of transportation network is complicated, aiming to satisfy the needs of society. The design process can be split into steps such as site investigation, route planning and design, earthwork calculation, pavement design, bridge and tunnel design, and soon. DTMs help in route planning and design earthwork calculation.

Designers make every effort to select a route passing through areas with stable geological conditions, with gentle slopes and small curves to minimize earthwork. Traditionally such work was done on contour maps. Nowadays DTMs are widely used for drawing plans, profiles (along the designed central line), and cross sections, for computing the volume of earthwork, for generating perspective views, and even for producing 3-D animation. As various routes are possible for a given project, the aim of the design is to find an optional route.

Other Applications

DTM can also be used for communication network planning. Problems such as dead angels and blind areas in site selection of the radio or television transmitting station can be computed.

DTM are also used for planning and landscape design, Visual Impact Analysis is applied to new designs that is the designs are superimposed on to a DTM to create a virtual landscape, which is visually analysed.

DTM Accuracy

Just as there is no such thing as an absolute accurate map, so the absolutely accurate terrain model does not exist. All digital terrain models will contain inaccuracies, to a greater or lesser extent depending on a number of interrelated factors. As with all mapping operations, the accuracy of the terrain model must be suited to the chosen or intended application.

The only 'true' situation in the context of mapping is the terrain surface itself, and since this condition of absolute accuracy cannot be attained by measurement, the accuracy of any field survey data, photogrammetric measurement or complete map can only be assessed by check comparisons with measurements made to a known higher order of accuracy.

If $v1$, $v2$.—— vn represent the residual at the n individual points, RMS error is one way of representing the accuracy.

$$RMS(M) = +/-\sqrt{\Sigma v^2/(n-1)}$$

Assuming the mean error to be zero and that there is a normal distribution, then 68.27% (approximately two-thirds) of the residual values will fall in the range $-M$ to $+M$. The term standard error is frequently employed in mapping to describe this expression of accuracy.

With respect in conventional mapping, there are two principle methods of recording and representing height information, both of which may be used as input data for DTMs.

(a)   Spot heights (regularly or irregularly distributed) and

(b)   Contours at a chosen vertical interval related to map purpose, scale and terrain slope.

In both cases, the accuracy has to be considered in terms of both plan position and height value. Spot heights are normally assessed or specified with respect to accuracy in terms of RMSE values related to plan and height. Contours are linear features, and their accuracy is not quite so simple to define as point-related data. The accuracy of contour lines lies in their fidelity to represent the minor (but often critical) irregularities of the terrain and morphological correctness of their shape. This relates to the detail of the contour rather than the general form.

Ref. Fig.1, the line A represents a highly accurate contour, such as might be determined by precise photogrammetric measurement, line B represents the much smoother from the same contour which might be the result of interpolation from a DTM. The 'Quantitative accuracy of B might well be quite high, but there is obviously qualitative inaccuracy and this is very often difficult to express in numerical terms.

Conclusion

Digital Terrain Modeling has become a discipline in itself. DTM have got wide range of applications. The accuracies and techniques of obtaining Digital Terrain models are evolving rapidly.   India 's CARTOSAT-1 is likely to meet increasing demand of users for a good quality DTM

Acknowledgement

The information presented in this paper is mainly sourced from two excellent books on digital terrain modeling by G. Petrie & Zhillin Li.

Spatial Perspectives for Understanding Landscape to Species Level Diversity

Murthy, M S R , Pujar G S and Giriraj, A

Forestry and Ecology Division,
National Remote Sensing Agency,
Hyderabad

Introduction

Attempt to derive ecological information from landscape level biodiversity to that at species level is an onerous one requiring nesting of information prevalent across a hierarchical scheme. But the need to evolve a comprehensive approach to handle the variability of diversity across scales is growing due to the range of hostile causes operating therewith (Turner, 1989; Riitters, et al 1997; Farina, 1998; Brokaw and Scheiner, 1989; Burnett et al, 1998 ; Nicholas et al. 1998;). Alteration of different units of natural resources in terms of their spatial and temporal contexts due to human induced factors has been alarming and requires involved sustainable solutions to counter the undesired progressions(Sala et al 2000). The global scale impacts brought about by remote factors or local factors operating at large expanse essentially demand synoptic observations whereas detailed characterization of local scale structures and processes would have to be addressed using additional sensing abilities. Consequent degradations prevalent almost cutting across the levels of development in hitherto healthy natural systems, have to be fully understood so that the remnants are conserved and the causes countered at appropriate level.

Approach to generate information base regarding diversity in spatial sense, can be two pronged, either building it from the existing knowledge regime of local scales or use synoptic abilities to build database at regional scale and downscale it to derive most detailed understanding. Either case would involve efficient telescoping of the information so that processes are modeled in full perspective and representation of key structural components considered. An attempt of such magnitude would require collative and collaborative paradigm to integrate as well as populate details at finer scale over the existing valid sets and then to choose optimal data layers for modeling coarse scale pattern.

Advanced Spatial Tools For Landscape Level Comprehension

Enhanced ability to map the Earth bound land covers comprising multiresolution sensors have facilitated much detailed view in a panoramic perspective . Coarse scale remote sensing systems provide synoptic view of the natural vegetation, with real time monitoring abilities around the year. Innovative observations of the Earth with such technology coupled with the realization of global impacts of remote anthropogenic actions confounded landscape view of the natural systems and have brought home the immediacy of the conservation priorities. Conceiving vegetation and allied components of land bound systems as landscapes; in tune with principles of landscape ecology was strengthened due to remote sensing consequently enabling quantitative approaches for holistic management. Coarse resolution imaging systems compiled for entire globe over a period of annual and decadal period revealed the status of the vegetation vis-à-vis biophysical juxtaposing having the anthropogenic dimension as the key agent. Moderate resolution sensors operating at around 30 – 60 mts pixel size enable delineation of vegetation types(using dominant vegetation reflectance), fragmentation and association with observable anthropogenic structures (Narendra and Gadgil, 1998; Udayalakshmi et al., 1998) which can be used for coupling with medium to coarse detail physiographic/terrain information. Information content at further level can be derived from very high resolution remote sensing images acquired from platforms set at low earth to polar orbits. The imaging provides near-meter to sub-meter detail, which would be rather sole information available at such a modest cost compared to aerial imaging. Structural details of the forest canopy regimes experiencing various degrees of disturbance and recovery can be conveniently demarcated for inclusion in to further modeling efforts.

Deducing species level diversity

Coarse resolution imaging with its high multitemporal frequency provides functional characterization of landscapes and helps to develop regional understanding of processes essential for maintaining and sustenance of variety of life forms. Incorporation of process understanding like annual net photosynthesis, primary productivity, phenological trends towards developing framework of biodiversity conservation can be critically important. Gradients of vulnerable portions of landscapes can be delineated using multicriteria approach involving such parameters, enabling focus of action at relevant biogeographic contexts. Global scale landscape parameters can be derived from coarse resolution sensors operating from kilometer to 250 meters pixel level. Data from AVHRR sensors has been employed for understanding forest fragmentation at global scale (Ritters et al., 2000) which provides regional prioritization of global level focus of action. MODIS data stream would also facilitate land cover assessment as well as land cover change understanding which acts as key input for modeling the disturbance at several edges. On the other hand its slew of biophysical parameters enhance the ecophysiological valuation of landscapes considerably at spatial dimension.

Necessity of downscaling the coarse scale database to implementation scale is quite high under current circumstances, as practical schemes/measures to reverse the erosion of biodiversity are needed urgently (Skole and Tucker, 1993). Sense of urgency is due to the fact that economies at either end of resource availability, strain natural resource in complex and unsustainable ways coupled with the looming danger of ubiquitous global warming. Delineation and analysis of species specific habitats vis-à-vis the biophysical variables assumes considerable significance as the activity at scale immediate to landscape, since their well being ensures the holistic survival of the components of biodiversity.

Multispectral remote sensing at fine resolution scale enabled by IRS P6 RESOURCESAT and precursors with LISS III sensor onboard offers potential to understand ecosystem wise community delineations. Datasets accrued over specific seasons can be used to understand the dominant vegetation responses coupled with sound field protocol so that overall sub-ecosystem variability is delineated. As the sensor offers sensing in SWIR region sensitive to canopy moisture and crown cover, scope, for instance, can be harnessed of mesic canopy patches as keystone resources. Such a value can be significant in tropical seasonal forest tracts especially in view of the global warming pushing habitability to extremes.

Spatial datasets for conservation zoning

Study of forest landscapes in terms of patch composition and configuration can generate important information (O'Neill et al. 1988; Peter and Goslee 2001) for conservation and ecorestoration purpose. Large-scale gradients in landscape heterogeneity can be related to broad-scale patterns in the environment. Due to sustained pressure small fragmented forest patches experience lesser density of population and increase in the risk of extinction (Farina 1998).

A study conducted for spatial characterization of evergreen forests of Western Ghats, Tamil Nadu, India - an ecological hotspot, using remote sensing and GIS based analysis in conjunction with ground based phytosociological data showed landscape characters. Evergreen forests in this part of Western Ghats are distributed in four distinct hill ranges namely Nilgiri, Anamalai, Palni and Tirunelveli having different topographic, bioclimatic, disturbance levels and are characterized for their uniqueness in terms of floristics. Vegetation type map was prepared using IRS LISS III satellite data and used to study the patch characters in terms of patch size, number, shape, porosity and land cover diversity. The phytosociological characters namely species richness, diversity, similarity and community assemblages were studied using ground data collected from 95 sample points of 0.1 ha size. Patch size and number revealed distinct intactness and disturbance levels in these four hill ranges.

Evergreen forests in Tirunelveli hills having 216.09 sq km are distributed in 306 patches and in Palni hills with 285 sq km forests are distributed in 1029 patches indicating large level of fragmentation. Land cover diversity indicating the spatial heterogeneity of land cover was very high in Nilgiri hills and low in Tirunelveli hills. The spatial analysis helped to delineate homogenous large patches of evergreen forest, which can be adopted for appropriate conservation strategies. A total of 342 tree species belonging to 4490 stems were evaluated for phytosociology. Only 15-28% of similarity in terms of species distribution was found across the hill ranges. Conjunctive analysis of patch characteristics and species distribution showed high species richness in less fragmented evergreen forests and vice versa. The study identified the areas of prioritization of Palni and Nilgiri hills in terms of eco-restoration and Anamalai and Tirunelveli hills for conservation based on patch and phytosociological characteristics.

Floristic Characterization And Species Habitat Relations

While medium resolution satellite datasets, in tandem with ancillary spatial and nonspatial data , can build information framework for local scale priority setting, very high resolution datasets capable of high temporal flexibility and across track pointing would provide images hitherto with limited and convoluted access. These images can be collected at far modest prices compared to aerial imaging campaign, without being perturbed by very high degree security clearance protocols as well as in multispectral mode compatible to medium resolution pixels. This enables synthesis of multiscale database in time and cost efficient fashion in turn helping a desired framework for data plugging from varied research/operational domains.

Such a high degree of detail available with their regional and global scale spatial alignment stacked in the information system, can be an ideal test bed to integrate local scale observations, manipulations, perceptions and aspirations in to it. Developing scenarios of utilization, exploration and conservation of resources can be accomplished by virtue of image based spatial modeling using GIS capabilities.

Floristic understanding of the regimes of diversity is paramount over the validity of the information generation and its end use by virtue of its ability to characterize the life form patterns and related resource allocation trends. Ability of this data to indicate several other bio-edaphic affinities makes it unique as a signature of ecosystem health. Current phytogeographic problems such as invasive species, climate change linked species vulnerability and range of similar situations do rely on precise definition of baseline vegetation composition, which in turn relies on floristics. Potential inherent in the remote sensing approach to define vegetation patches better in terms of variability in a spatially explicit manner has strenghthened the way sampling schemes are decided to understand the land cover health and changes.

Very high resolution image based information can act as the key link in upscaling phytosociological understanding. Currently the medium resolution dataset permits coarse level spatialisation of this and limits a sound site scale appreciation at times. The upscaling can be spectrally and spatially explicit to yield a hierarchically robust teleology in managing diversity patterns.

Quantitative floristic inventories based on one-hectare plots have been used in recent years to characterize forest vegetation throughout the tropics (Parthasarathy 1999, 2001; Amarnath et al, 2003). Quantitative floristic inventory was conducted using three hectare plots showed high species richness, diversity and endemism. Species packing can be more reliably understood for fine scale understanding using plots of higher spatial coverage. Based on the species dominance, Kakachi evergreen forest can be designated as *Cullinea exarillata – Palaquium ellipticum – Aglaia bourdillonii* series. Analysis of the floristic data in Kakachi RF allows the detection of five main mesoscale floristic units.

With the present conservative practices and increasing resident population due to tea, coffee and cardamom plantations activities in the core areas of KMTR, the forest type can be expected to be disturbed evergreen system or fragmentation of climax forest or undergo change in type and kind of species in the near decade. The tribal / local people also exploit the natural resources for their regular need. Thus, management of medium elevated evergreen forests must necessarily depend on knowledge of recognizable community types and their environmental variables. These methods of data collection and analysis offer the environmental managers to sustain and preserve such communities. This approach comprising of intrinsic data associated with landscape characteristics can be utilized by the ecologists to predict endemic habitat zonations.

Modeling Species Occurrence And Validation
Intensive field extrapolation for species representation and characteristic habitat
High degree of endemism marks the key importance of the tropical flora and fauna. The unperturbed evolutionary time vectors in absence of glaciations have facilitated large number of co-evolved species-species, species-habitat situations. Slightest offsetting of links in such co-evolved trophic chains can seriously hamper persistence of species and assemblages. Geoinformatic tools armed with precise geo-locating devices as well as fine scale bio-physical boundaries are potent to facilitate a high precision mapping of particular species. The significance of species can be due to its utilitarian, functional or supportive values. The characterization of habitat can be accomplished by virtue of high to medium resolution satellite images integrated with global level datasets of meterological parameters and terrain as well as linking any non-spatial ancillary data in GIS.

Modeling the spread as fundamental niche
Population spread of the targeted species can subsequently be modeled on the basis of spatial logic. Field data configured for spatial 2D alignment can be subjected to spatial autocorrelation tools of geostatistics to provide possible contours of spread of the speices per se or its habitat. The method might have higher assumptive echelon since predicted contours may mostly obey linear directive due to mathematical rigidity at first hand , which

may in practice be refined for nonlinearities due to biological compulsions once the validation is effected. Geostatistics approach may model the biophysical spread with greater degree of confidence. To overcome the problems inherent in such an approach, methods taking care of spatially corresponding parameters are devised, which would act according to the spatial model which takes care of nonlinearity inherent in biological systems. In general Geographic information systems can be resorted to generate such a spatially explicit spread of the species. However, handling large number of empirical as well as quantitative, genetic algorithm based tree classifiers have shown considerable promise. The ability to create possible set of rules by writing off the totally irrelevant ones using chromosome logic has opened immense possibilities to predict species niches across range of scales.

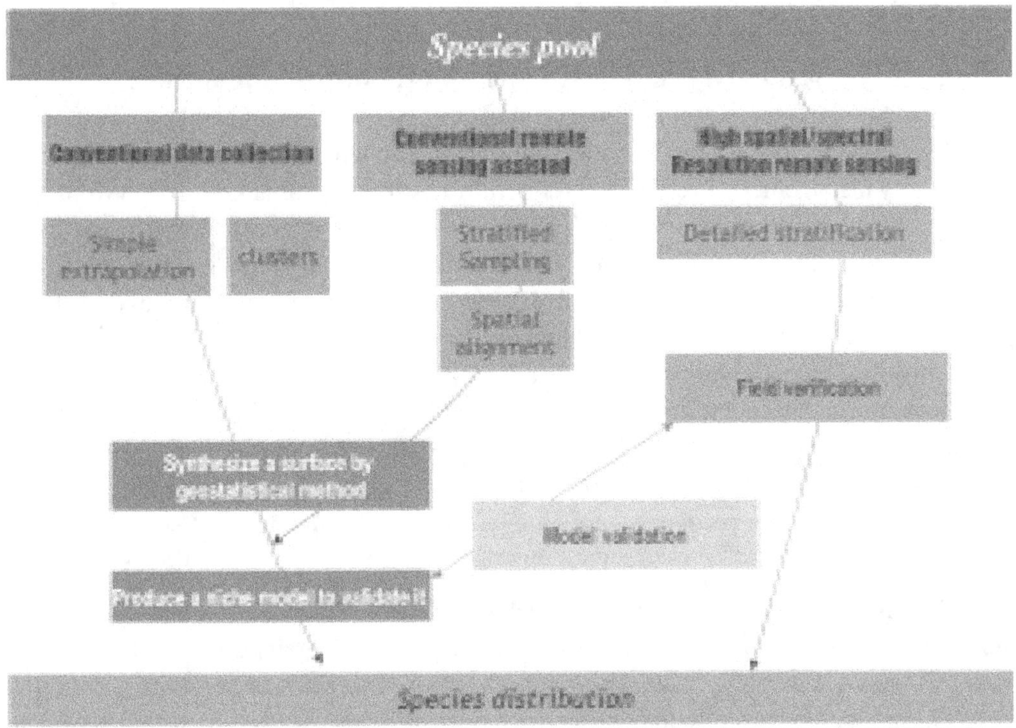

Fig. 1 Understanding species presence- Spatial and point based approaches

GARP (Genetic Algorithm for Rules set Prediction) can be used to identify regional limitations and managed habitat associations via the fine-grained land-use/land-cover map and other bioclimatic variables. The environment uses GA to arrive at different set of rules as alternatives with spatial inputs , for modeling the fundamental niche of a given species. Whereas field observation are obviously from a realized niche, the model points to the habitability of the individual, which should in turn be validated. GARP models of species' distributions visualized in geographic space are expected to provide a guide for future sampling. Modeling of RET (Rare, Endangered and Threatened) species using GARP can identify the probable areas of their occurrence for conservation purpose. Currently GARP is applied for modeling an endemic and threatened plant species - Aglaia bourdillonii in Western Ghats to identify extent of ecological niche in geographical space. It seems that Aglaia bourdillonii, a threatened and endemic species in Southern Western Ghats, requires very specialized habitat conditions as well as specific associations. The distribution of Aglaia bourdillonii is also restricted by terrain and topography. In fact in conservation of Aglaia bourdillonii it is essential that the association of Aglaia bourdillonii, Cullenia exarillata and Palaquium ellipticum is conserved encouraging niche diversity to increase both in species assemblage as well as in space.

Inculcation Of Values Of Biodiversity In Land Resource Management

Majority of the demographic situations capable of bringing about landuse changes especially in biodiversity hotspots need to be handled for a win-win situation with greater reliance on participatory management. Prerequisite for a fruitful, strong sustainable solution would be a more non-language/non-text oriented depiction of solution, rather than verbal or textual communication. Since several stakeholders, hitherto unexposed to modern living level can easily associate themselves with land parcels more than any explanatory protocol. Geoinformatic rendering of diversity related data as well as scenarios can be a very potential opportunity to associate various stakeholders with their respective scales of operation. Multiresolution database comprising of information from landscape diversity to species level diversity should be part of such a plan, so that threats and opportunities with regard to local and regional diversity should be properly addressed. Such a spatial and holistic scheme would be able to ensure recovering balance between demands on the system against the demand pattern.

Coarse resolution datasets along with regional modeling of anthropogenic demands in space and time can reveal the scope of staggered handling of biodiversity erosion. Diagnosis of pressures vis-à-vis diversity using geographic analysis especially in terms of proximities, market juxtaposition, harvest channels, terrain amenability for access, sustenance demands etc would demarcate zones of immediate concerns. Based on such a scheme, participations may be initiated to induce required changes in land management. Objectives of dialogue can address :

1. Significance of landscape level quality to the land manager
2. Impact of networked action of natural resource managers/entrepreneurs on the upkeep of landscape health

3.   Achieving conservation of individual species especially endemics,

4.   Highlight the incentives accruing of the conservation for base level land managers (for instance, tourism, medicinal plant products, and improved minor forest product yield by bare foot tree breeding etc.)

Significance of regional level information network

Magnitude of land cover changes controlling depletion of forested landscapes are varied across global contexts. Variability of degraded fragments altering forest landscapes mainly in terms of permanence, patch size and shape, demand characterization across scales. Complexities inherent in tropical forest ecosystems and their derived natural resource systems create a range of landscape configurations comprising of highly interspersed patch configurations due to fragmentation. Major causes for fragmentation lie in anthropogenic sources. Coarse resolution global scale databases attempting to characterize such a context may not delineate the intricate spatial pattern in required detail. Fine scale resolution based information becomes imperative to derive database to cater for diverse yet local specific decision making for sustainable natural resource management.

The interests with regard to diversity have to be understood based on ecological as well as economical perspectives. While many of the flagship species fall prey to illegal trade endangering sustainability of the complete trophic systems, landuse changes facilitate a low key but high intensity erosion of the flora and fauna. The genuine requirement of products and services of diversity hence may not be considered bonafide often in spite of its significance. Considering the local market level mechanisms of harvest, collection and trade of the product it may be essential to consider the genuine requirement and educate the stakeholders for sustainable handling.

Diverse stakes ranging from sustenance to corporate interests need to be scrutinized in view of sustainability. Degree of tangibility of the species utility varies across hierarchy of scale. While most of ecosystem values of diversity are generally realized for instance either as value of a molecule or an infestation menace, impact on the economy, health and other social systems are increasingly acknowledged seriously. Intangible values like microclimate moderation, wilderness value, recreation, conservation of flagship species need to be internalized in the valuation system and may require strong spatial component to strike a better outreach. Conflict of stakes (Fig. 2) can get a considerable leverage from a trans-scale understanding of the issues of diversity from landscape to species. The inherent nature of the threat and supportive value derived from human presence in various contexts of conservation, can precisely be aligned in spatial scheme so that visualization, compilation, analysis and solving the ecosystem issues is accomplished. As the impacts of any human activity have ability operate at various scales, a comprehensive understanding of hierarchy of organization of biotic entities is required. Hence, landscape to species level understanding would have a critical end use application in establishing a conservation schemes with technologically assisted conflict moderation.

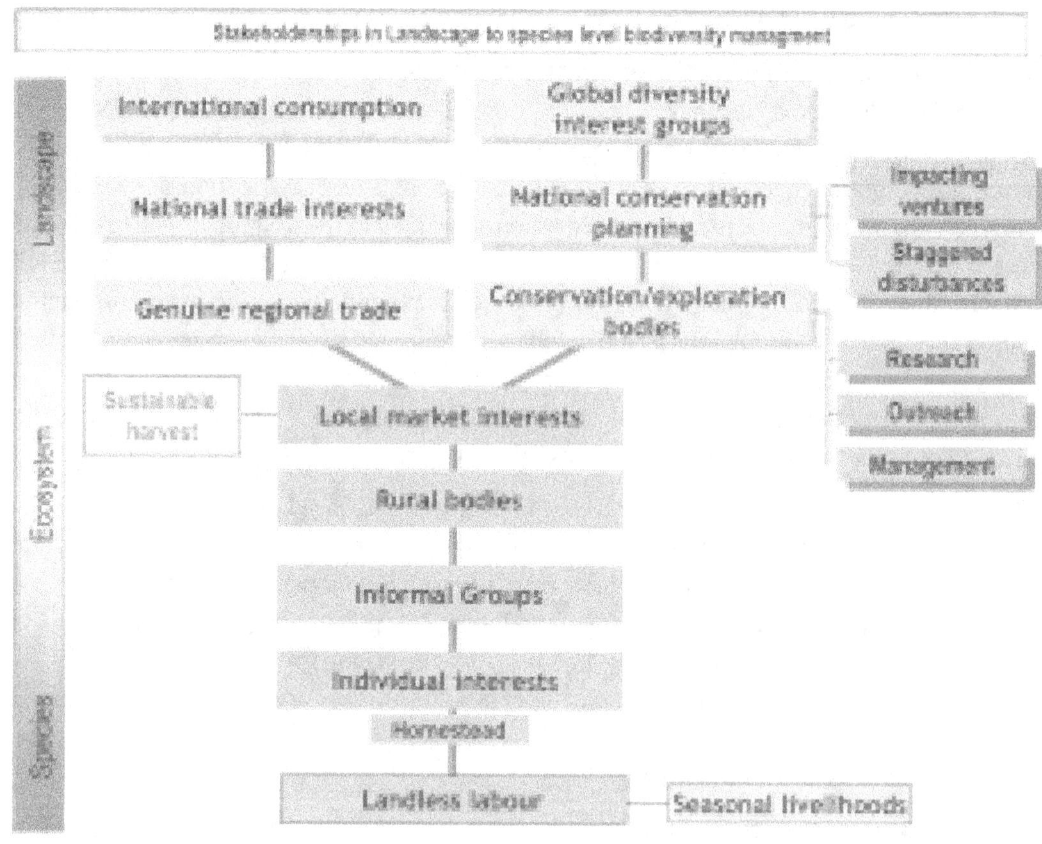

Conclusion

Given the significance of the variability of the life and its variety , conservation paradigm should consider a seamlessly integrating information base for holistic action. Deriving such a base can be a paramount prerogative of spatial information domain wherein synergy of suite of multispectral and multiresolution sensors on board satellite and aerial platforms is available. Enriching the information using the ground based data to validate as well as to integrate with the satellite derived datasets can facilitate a comprehensive approach in this regard. Database can be further put to advanced spatial modeling to derive species distribution as defined by its biophysical habitat traits. Derived spatial distribution suitably integrated with coarser scale information can be used for resolving the stakeholder interests to achieve conservation, by geospatial query, visualization and analysis.

References

Amarnath, G., Murthy, M.S.R., Britto, S.J.,Rajashekhar, G and Dutt, C.B.S. 2003. Diagnostic Analysis of conservation zones using remote sensing and GIS techniques in wet evergreen forests of the Western Ghats – A ecological hotspot, Tamilnadu, India. Biodiversity and Conservation. 12:2331-2359.

Brokaw N.V.L. and Scheiner S.M. 1989. Species composition in gaps and structure of a tropical forest. Ecology 70: 538-541.

Burnett M.R., August P.V., Brown J.H. JR and Killingbeck K.T. 1998. The influence of geomorphological heterogeneity on biodiversity. I. Patch-scale perspective. Conservation Biology 12: 363-370.

Farina, A 1998. Principles and Methods in Landscape Ecology. Chapman and Hall. London

Levin, S. A. 1992. The problem of pattern and scale in ecology. Ecology 73:1943-1967.

Nagendra, H and Gadgil, M. 1998. Linking regional and landscape scales for assessing biodiversity : A case study from Western Ghats. Current Science. 75:264-271

Nichols W.F., Killingbeck K.T. and August P.V. 1998. The influence of geomorphological heterogeneity on biodiversity. II. A landscape perspective. Conservation Biology 12: 371-379.

O'Neill R.V., Krummel J.R, Gardner R.H., Sugihara G., Jackson B., De Angelis D.L et al 1988. Indices of landscape pattern. Landscape Ecology . 1:153:162

Parthasarathy, N .1999. Tree Diveristy and distribution in undisturbed and human-impacted sites of tropical wet evergreen forest in southern Western Ghats, India. Biodiveristy and Conservation. 8:1365-1381

Peters, D.P.C and Goslee, S.C. 2001. Landscape diversity. In: Levin S.A(ed). Encyclopedia of Biodiversity. Vol 3. Academic Press. New York. Pp 64-658

Riitters, K. H., R. V. O'Neill, and K. B. Jones. 1997. Assessing habitat suitability at multiple scales: a landscape-level approach. Biological Conservation 81:191-202.

Riitters, K., J. Wickham, R. O'Neill, B. Jones, and E. Smith. 2000. Global-scale patterns of forest fragmentation. Conservation Ecology 4(2): 3. [online] URL: http://www.consecol.org/vol4/iss2/art3/

Roy P.S., Porwal M.C. and Sharma L. 2001. Mapping of Hippophae rhamnoides Linn. in the adjoining areas of Kaza in Lahul and Spiti using remote sensing and GIS. Current Science 80: 1107-1111.

Skole, D., and C. Tucker. 1993. Tropical deforestation and habitat fragmentation in the Amazon: Satellite data from 1978 to 1988. Science 260:1905-1910.

Sala, O.E Chapin, S.F., Armesto, J.J., Berlow, E et al. 2000. Global biodiversity scenarios for the year 2100. Science. 287. 1770-1774.

Turner, M.G. 1989. Landscape ecology: the effect of pattern on process . Annual Review of Ecology and Systematics. 20:171-197

Udayalakshmi V,Murthy, M.S.R.and Dutt , C.B.S . 1998. Efficient forest resources manamgement through GIS and remote sensing . Current Science. 75:272:282

The Role of Geoinforamtics in Urban Planning

Subhan K. Pathan

Head, Land use Planning and Photogrammetry Division
Space Applications Centre (ISRO) Ahmedabad

Introduction

There are 5161 towns and cities in the country (Census 2001) with an urban population of 285 million. While the share of urban population, employment and contribution to the urban economy has continued to grow, towns and cities suffer from over crowding, congestion, slums and lack of civic services, etc. undermining the quality of the environment. This situation poses a challenge for planning and management of urban areas in the light of liberalization and globalisation. What is required is systematic compilation of spatial and attribute data and information related to all urban settlements in order to plan and manage development effectively. Of late, there have been tremendous advancements in data base management systems with the aid of computer technology and communication systems but on the other hand a lot of data particularly at local levels remains unorganised and inaccessible. Standardization of data systems and structures are still in the rudimentary stage. The 74th Constitution Amendment Act has laid down provisions for spatial economic and environmental planning in the planning system at all levels, be it a Nagar Panchayat, Municipal Council or Municipal Corporation. These functions have been assigned to urban local bodies marking a radical shift from a **'top down'** to a **'bottom up"** approach.

With increasing access to modern data sources and automation in processing and analysing data, most organizations have embarked into application of latest technology i.e. 'Geoinformatics'. 'Geoinformatics' Technology is a new discipline integrating elements of various disciplines dealing with geographic data i.e. data linked to a particular location on the surface of the earth. Therefore, 'Geoinformatics' is the science and technology for collection, management, analysis and presentation of geographic and other spatially defined data. It deals with information about objects, phenomena and processes on and in the earth, such as physical environment, natural and man made resources, their use and changes through time. Therefore it includes Surveying, Remote Sensing, GIS, GPS, Photogrammetry, Cartography, Geography, Computer Science and Statistics. Thus, it focuses not only theory and practical design and development of geographic information systems but also the analytical methods that are characteristically required in a variety of applications. Hence, Geoinformatics Technology has stimulated the entire scientific and administrative communities in the country by providing the theoretical knowledge and practical skills to create geographic information systems and to develop a scientific approach to the fundamental issues in respect of the urban development and use of these systems. Space Applications Centre (ISRO), has taken up the lead way back in 1981, to create Geographical information systems to cater the needs of various Town Planning Departments, Urban Development Authorities and Municipal Corporations. A number of projects have been jointly carried out with these agencies to develop methodologies for urban sprawl, urban land use, urban land use suitability analysis, urban environmental sensitivity analysis etc. employing various Geoinformatics techniques. These methodologies led to the formation of "Urban Information Systems" in the country for urban hierarchical planning and management. These technologies offer powerful analysis tools, which provide multiple scenarios for decision support. All these factors bring out the urgency and need for sustainable urban planning at all levels. Realising the potential of Remote Sensing and GIS technologies, The Planning Commission, Government of India is now of the view that all components related to spatial and attribute data be integrated in one system called the "National Urban Information System (NUIS)" to cater the needs of different hierarchical levels of urban planning. This paper highlights the potential of Remote Sensing and GIS techniques with some specific case studies.

Urban Planning And Information Needs

Planning is a technique of resources allocation for the convenience, safety and pleasure of the people. Planning includes town planning, city planning, regional planning, national planning and global planning. At the town level, planning pays a detailed attention to matters such as housing, providing schools, shops, and rural industrial estates, as also the communications with nearby villages. At the city and metropolitan level, the planning process comes across more complex problems, such as industrial development, movement of people from their residence to work space, transportation of goods, residential accommodation, utilities and services, quality of life, environmental problems, zoning of land and others. At the regional level, the planning seeks to serve the interests of several communities, individually or in relation to one another taking into account the resources available and the optimum utilisation of these resources to the best advantage of all the communities.

The planning and formulation of policies and programs for urban development have been very much handicapped by the non-availability of data on physical inputs, socio-economic status, information maps and proper documentation. The maps, which are available for physical planning, are outdated by 30 to 50 years. Up-to-date mapping of a city using conventional means would involve considerable amount of time, funds and manpower. Moreover, by the time the map is produced, the changes would have become significant. This has led the planners and administrators to search for modern tools for data acquisition. With the advent of space systems a new dimension has been added to remote sensing.

Therefore, the planning procedure for the preparation of a development plan has changed through time. This is primarily due to the dimensionality of input data has increased tremendously through time. Thus the preparation of 'Development or Master Plan' methodology has been modified to incorporate more and more parameters. It is very difficult to collect and integrate various types of data mentioned above using conventional methods. Hence, it has become necessary to depend on new type of surveys/methods/techniques, which can provide comprehensive and exhaustive information on various aspects of natural resources in the urban environment. Secondly, there is a need to integrate this information to study the combined effect of these parameters and their interrelationships in the urban environment. In recent times, Remote Sensing (RS) technology has proved to meet these data requirements reliably, timely, economically, accurately and periodically. GIS technology has proved to integrate both spatial and non-spatial data efficiently and helps to create alternate planning scenarios by studying the interrelationships between these parameters for decision making (Revised Development Plan of AUDA, 1997).

Remote Sensing

Remote Sensing is a technique of collecting information about objects and features without there being any physical contact with them. The instruments known as sensors are placed on aircraft or spacecraft platforms to survey the earth's surface. These sensors quantitatively register the reflectivity and emissivity of objects in various bands of electromagnetic spectrum. The images produced by these sensors are used for finding out the natural resources on the earth's surface.

Space based remote sensing is being currently used for various applications of national relevance. The data is utilised to prepare thematic maps and generate information on land use patterns, geomorphology, ground water, geology, urban sprawl etc.

Though the satellite data is available from 1972 onwards, Indian Remote Sensing Satellites viz. IRS-1C/1D satellites have opened a new chapter in the history of satellite remote sensing data utilisation for urban planning especially in the area of physical planning of the city and its environs (Pathan et al., 1991 and Dhinwa et al., 1994). The IRS-1C satellite launched on December 30, 1995 has three types of payloads viz. Panchromatic (PAN), LISS-III and WiFS. Panchromatic camera collects the data in both nadir and stereo mode in the spectral region of .5 to .75 μm. This camera has a spatial resolution of about 5.8 m and it also provides stereo coverage across the track. The scene size is 23.5 Km by 23.5 km with a revisit capability of 5 days. It has a radiometric resolution of 6 bits. LISS-III camera has a spatial resolution of about 23 m with four spectral bands in visible, near infrared and shortwave infrared regions. It covers an area of about 140 Km by 140 Km with a revisit capability of 24 days.

With the advent of IRS-1C/IRS-1D satellite high spatial resolution data (Panchromatic – 5.8 m, LISS-III – 23 m), it has become possible to prepare urban land use maps at 1:12,500 scale depicting level-II urban land use classes and in some cases even Level-III urban land uses (**Figure-1**). The standard urban land use classification system designed in consultation with various Town Planning and Valuation departments (TPVD's), Development Authorities and Metropolitan Region Development Authorities is presented in **Appendix-I.** Most of the information mentioned in this can be easily obtained using CARTOSAT data of 2.5 m spatial resolution and 1m spatial resolution of IKONOS and QUICK BIRD and will be possible to prepare urban land use maps at the scales required by most of TPVD's and Development Authorities in the country. The details extractable with various sensors with respect to urban information needs are presented in **Appendix-II**.

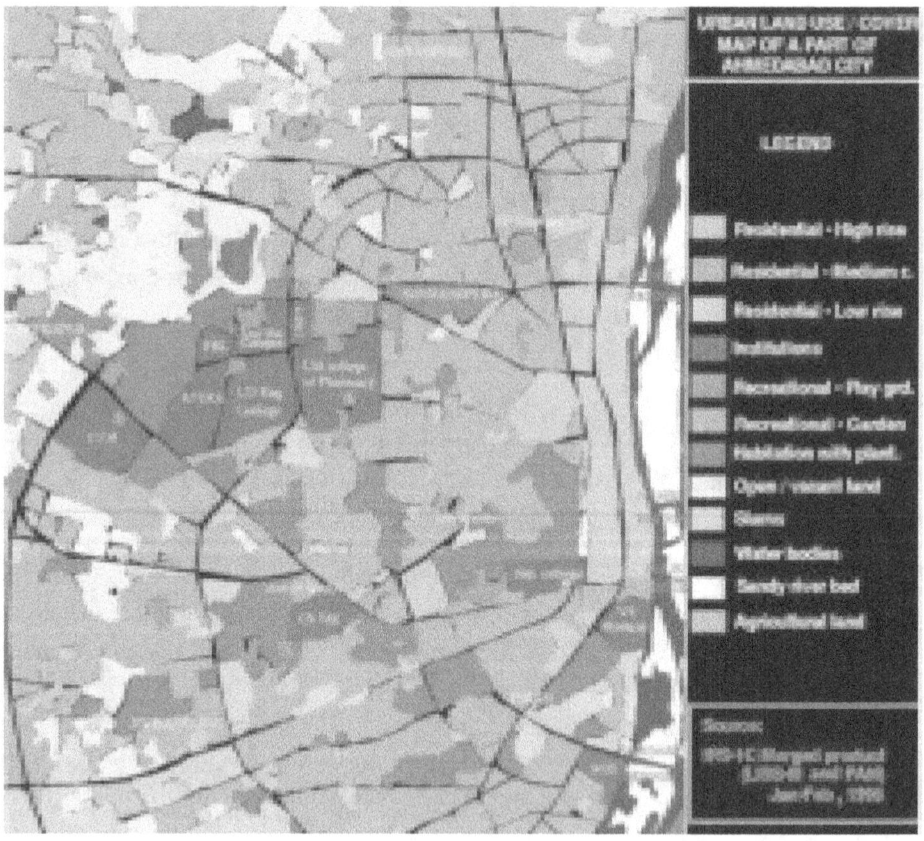

Figure-1 : Urban land use map of a part of Ahmedabad

Appendix-I : Urban Land Use/ Land Cover Classification System

URB-CODE	LEVEL - I	LEVEL - II	LEVEL - III	LEVEL – IV
01-00-00-00	Built Up			
01-01-00-00		Built Up (Urban)	Residential	
01-01-01-00			High density residential	
01-01-01-01				High rise-Apts/Flats (30 m+)
01-01-01-02				Medium rise Apts/Flats (15-30 m)
01-01-01-03				Low rise Apts/Flats (< 15 m)
01-01-01-04				Low rise Row houses
01-01-01-05				Low rise Group Houses / Tenaments
01-01-01-06				Slums/Clusters
01-01-01-07				Others
01-01-02-00			Medium density residential	
01-01-02-01				High rise-Apts/Flats (30 m+)
01-01-02-02				Medium rise Apts/Flats (15-30 m)
01-01-02-03				Low rise Apts/Flats (< 15 m)
01-01-02-04				Low rise Row houses
01-01-02-05				Low rise Group Houses / Tenaments
01-01-02-06				Slums/Clusters
01-01-02-07				Others
01-01-03-00			Low density residential	
01-01-03-01				High rise-Apts/Flats (30 m+)
01-01-03-02				Medium rise Apts/Flats (15-30 m)
01-01-03-03				Low rise Apts/Flats (< 15 m)
01-01-03-04				Low rise Row houses
01-01-03-05				Low rise Group Houses / Tenaments
01-01-03-06				Slums/Clusters
01-01-03-07				Others
01-01-04-00			Industrial	
01-01-04-01				Service Industry
01-01-04-02				Light Industry
01-01-04-03				Extensive Industry
01-01-04-04				Heavy Industry
01-01-04-05				Hazardous industry (chemical/pharmaceutical)
01-01-04-06				Others
01-01-05-00			Mixed Built Up area	
01-01-06-00			Recreational	
01-01-06-01				Parks/Gardens
01-01-06-02				Stadium
01-01-06-03				Playgrounds
01-01-06-04				Golf Course/Race course
01-01-06-05				Zoological parks / Botanical gardens
01-01-06-06				Historical monuments

URB-CODE	LEVEL - I	LEVEL - II	LEVEL - III	LEVEL – IV
01-01-06-08				Exhibition / Function hall
01-01-06-09				Swimming pool/Gymnasium
01-01-06-10				Cinema halls / Theatres
01-01-06-11				Others
01-01-07-00			Public and semipublic	
01-01-07-01				Educational Institutes
01-01-07-02				Cantonments
01-01-07-03				Hospitals
01-01-07-04				Cremation/buried ground
01-01-07-05				Social and Cultural center
01-01-07-06				Religious places
01-01-07-07				Government Office
01-01-07-08				Petrol/Gas filling stations
01-01-07-09				Police Station
01-01-07-10				Fire station
01-01-07-11				Rest office
01-01-07-12				Electric sub-station
01-01-07-13				Jail
01-01-07-14				Bank
01-01-07-15				Others
01-01-08-00			Communications	
01-01-08-01				Post office
01-01-08-02				Telephone exchange
01-01-08-03				Telegraph office
01-01-08-04				Radio/TV station
01-01-08-05				Radar station
01-01-08-06				Pylon
01-01-08-07				Others
01-01-09-00			Public Utilities & facility	
01-01-09-01				Water Treatment Plant
01-01-09-02				Sanitary Landfill
01-01-09-03				Electric Power plant
01-01-09-04				Sewerage Treatment Plant
01-01-09-05				Overhead tanks
01-01-09-06				G L R
01-01-09-07				Others
01-01-10-00			Commercial	
01-01-10-01				Retail and General Business
01-01-10-02				C.B.D/Sub C.B.D.
01-01-10-03				Community center
01-01-10-04				Wholesales and Warehousing
01-01-10-05				Local shopping centers
01-01-10-06				Hotel/Restaurant
01-01-10-07				Mall
01-01-10-08				Parking lots
01-01-10-09				Market yard
01-01-10-10				Others
01-01-11-00			Transportation	
01-01-11-01				Bus stands
01-01-11-02				Railway stations
01-01-11-03				Air ports

URB-CODE	LEVEL - I	LEVEL - II	LEVEL - III	LEVEL – IV
01-01-11-07				Railway tracks
01-01-11-08				Air strips
01-01-11-09				Groynes / break water / Jetties
01-01-11-10				Others
01-01-12-00			Reclaimed land	
01-01-13-00			Vacant land	
01-01-13-01				Lay out
01-01-13-02				Under Construction
01-01-14-00			Vegetated Area	
01-02-00-00		Built Up (Rural)		
02-00-00-00	Agriculture			
03-00-00-00	Forest			
04-00-00-00	Grassland /grazing land			
05-00-00-00	Wastelands			
06-00-00-00	Wetlands			
07-00-00-00	Water bodies			
08-00-00-00	Others			

Structure of Look Up Table (LUT)

Field Name	Field Type	Key field (Y/N)	Remarks
LU-CODE	8,8,C	Y	Feature code
Discr-L1	30,30,C	N	Level-I Classes
Discr-L2	30,30,C	N	Level-II Classes
Discr-L3	30,30,C	N	Level-III Classes
Discr-L4	30,30,C	N	Level-IV Classes

Appendix-II : Remote Sensing versus Urban Information Needs- On the basis of studies conducted on different cities using RS data, the following comparative table has been prepared. This table provides the information needs that can be met using various types of RS data.

S.No	Information need	Level of Planning	Unit size	Scale of mapping	Remote Sensing capability		
					(a)	(b)	(c)
A).	*Physical setting*						
1	Topographic features	Macro	Metropolitan Region	1:250,000	*		
2	Geological base and Minerals				*		
3	Soil associations				*		
4	Drainage pattern				*		
5	Land use/cover				*		
6	Surface and ground Water conditions				*		
7	Climatic conditions				$		
B).	*Economic base*						
1	Agriculture	Macro	Metropolitan Region	1:250,000	*		
2	Forestry				*		
3	Industry				@		
4	Construction materials				@		
5	Commercial					*	
C).	*Transport and Communications*						
1	Road/rail network	Macro	Metropolitan Region	1:250,000	*		
2	Condition of roads					*	
3	Canal network				*		
4	Traffic volumes						*

(a) = Amenable to satellite data with some ground truth collection, (b) = Amenable to aerial data with ground truth collection (c) = Not amenable to remote sensing

* Possible   @ Possible with high spatial resolution data $ Possible with meteorological satellite data

Remote Sensing Versus Urban Information Needs

S.No	Information need	Level of Planning	Unit size	Scale of mapping	Remote Sensing capability		
					(a)	(b)	(c)
Meso level planning : Preparation of Urban Development Plan							
1	Built-up land	Meso	Development Authority area	1:50,000	*		
2	Residential				*		
3	Commercial						*
4	Industrial				*		
5	Recreational				*		
6	Transport & Communications				*		
7	Public & Semi public				@		
8	*Open spaces*				@	*	
9	*Wastelands*				@	*	
10	*Water bodies*				@	*	
11	*Agricultural area*				@	*	
12	*Landforms*				@	*	
13	*Geology*				@	*	
14	*Natural hazards*				@	*	
15	*Construction materials*				@	*	
16	*Drainage*				@	*	
17	*Condition of roads*						*
18	*Traffic volumes*						*
19	*Canal network*					*	
20	*Population*						*

(a) = Amenable to satellite data with some ground truth collection,   (b) = Amenable to aerial data with ground truth collection

(c) = Not amenable to remote sensing

* Possible   @ Possible with high spatial resolution data

Remote Sensing Versus Urban Information Needs

S.No	Information need	Level of Planning	Unit size	Scale of mapping	Remote Sensing capability		
					(a)	(b)	(c)
A).	Land use/cover						
1	Residential – High rise	Mciro	Urban Complex Area / Ward	1:7920 and larger	@	*	
2	Residential – Medium rise				@	*	
3	Residential – Low rise				@	*	
4	Industry – Heavy				@	*	
5	Industry – Light					*	
6	Commercial – large					*	*
7	Commercial - small						*
8	*Roads – Major*				@	*	
9	*Roads – Minor*				@	*	
10	*Roads – Kutcha*				@	*	
11	*Railways – Broad gauge*				@	*	
12	*Railways – Meter gauge*				@	*	
13	*Runways*				*	*	
14	*Parks/Gardens/playgrounds*				*	*	
15	*Open spaces*				*	*	
16	*Govt. offices*				@	*	
17	*Godowns*				@	*	
18	*Slum areas*				@	*	
19	*Sewage network*						*
20	*Rivers/streams*				*	*	
21	*Lakes/ponds*				*	*	
22	*Terrain characteristics*				*	*	

(a) = Amenable to satellite data with some ground truth collection,   (b) = Amenable to aerial data with ground truth collection

(c) = Not amenable to remote sensing

* Possible   @ Possible with high spatial resolution data

Remote Sensing Versus Urban Information Needs

S.No	Information need	Level of Planning	Unit size	Scale of mapping	Remote Sensing capability		
					(a)	(b)	(c)
B) Housing characteristics							
1	Quality of housing	Micro	Urban Complex Area / ward	1:7920 and larger	@	*	
2	Housing density				@	*	
3	Type and structure of house				@	*	
4	Height of structure				@	*	
5	Dwelling unit size				@	*	
6	*Occupancy rate*						*
7	*Tenure status*						*
8	*Slum typology*				@	*	
C) Population characteristics							
1	Population density	Micro	Urban Complex Area / Ward	1:7920 and larger	@	*	
2	Age/sex structure						*
3	Occupational pattern						*
4	Literacy						*
5	Migration						*

(a) = Amenable to satellite data with some ground truth collection,   (b) = Amenable to aerial data with ground truth collection

(c) = Not amenable to remote sensing

* Possible   @ Possible with high spatial resolution data

Geographic Information System (GIS)

Optimal management of urban resources requires the attention for balancing the needs against the possibility of endangering the environment irreversibly. Such a system primarily requires systematic, detailed, reliable and accurate, timely information on the extent and spatial distribution of various natural resources, socio-economic, demographic patterns and cultural structures of the inhabitants. The data collected on different aspects of the natural resources has to be translated into useful information and converted into user defined formats. Lastly, there will be a need to aggregate this information according to administrative and natural resource units. The experiences gained regarding the existing natural resource information system in different fields of development clearly bring out the fact of several short comings in regard to acquisition of statistics, processing, generation of graphic outputs and their storing. In fact, these short coming acts as serious predicament for efficient and meaningful planning including implementation of programmes and monitoring of development. It is in this context GIS will play a major role by providing linkage between the information domain and the technologies available for natural resources development and management. Hence, GIS is a particular form of Information System that is applied to geographical data. An Information System is a set of processes, executed on raw data, to produce information which will be useful in decision-making. Therefore an information system must have a full range of functions to achieve its purpose, including observation, measurement, description, explanation, forecasting and decision-making. The integration of spatial information for the natural resources development has been done using manual methods in the past. But with the increase in the volume and dimensionality of data sets, it has become essential to use automated GIS to meet the demands in natural resources development. Use of an automated system has become necessary as the data are maintained in a physically compact format (i.e. magnetic media), data can be retrieved with much greater speed, and various computerised tools allow a variety of manipulations. Hence, GIS is defined as "An automated tool useful to capture, storage, retrieval and manipulation, display and querying of both spatial and non-spatial data to generate various planning scenarios for decision making" (GIS user manuals, 1989). Thus GIS is a tool able to answer the location, condition, trends, patterns and modeling. Integrated analysis related to urban land use suitability analysis (based upon various physical parameters), urban land use zoning and preparation of Perspective plan can be easily carried out with the help of both RS and GIS techniques.

Urban studies conducted using RS and GIS techniques

A number of pilot studies have been carried out to demonstrate the utility of RS and GIS technology in urban planning with a specific reference to the preparation of Master Plans and Regional Plans in the country. Glimpses of these studies are presented in this paper.

Macro level Urban Information System – A GIS case study of Mumbai Metropolitan Region

This is joint study carried out in collaboration with Mumbai Metropolitan Region Development Authority (MMRDA), Mumbai for the preparation of 'Regional Plan' covering an area of about 4350 sq km (Mukund Rao, 1992 and Pathan et al., 1992). The objectives the project were i) Design and organisation of a spatial information system for BMR on 1:250,000 scale and to serve as a macro-level data base for urban planning, ii) Demonstrate integrated analysis of spatial/non-spatial data for specific urban planning problems such as a) suitability assessment, b) environmental condition assessment, c) growth profile analysis, d) land use change analysis and e) routing analysis.

Resources inventory related to urban land use mapping, urban sprawl, hydrogeomorphology, groundwater prospects, flood and erosion hazards etc. has been done with the help of Remote Sensing data employing hybrid (visual and digital) interpretation techniques. Suitability assessment for urban development has been carried out based on a comprehensive model incorporating a multi-parameter data set employing GIS techniques. Environmental condition assessment has been done by the integration of the pollution data, land parameters etc. in GIS environment. Growth profile analysis has been carried out by establishing a relationship between the population and the physical extent of urban areas. This relationship was used for estimating the physical extent of the urban built-up land for the year 2001. Land use change analysis has been done based on a multi-data land use integration and identification of the change areas as a matrix in GIS. In addition to this, Routing analysis for the proposed airport in MMR was also carried out based on an evaluation of travel-distance/shortest-path and the travel times on the existing road network and a network including proposed freeways.

The derived outputs from this technology helped the planners to prepare the sustainable 'Regional Plan of MMR'. For example, the 'Urban land use suitability map' helped the planners in identifying the areas suitable for construction purpose and the areas to be conserved under greenery. Environmental condition assessment map helped them in allocating the best areas for urban development. The changes in land use pattern through time helped the identification of newer areas for urban development as well as the reasons responsible for different changes. Similarly, the urban sprawl map paved the way in identifying the direction and rate of growth of various urban areas in MMR. Routing analysis helped the planners to optimise routes for the proposed airport in Alibaug area.

The regional plan prepared on the basis of above inputs has been integrated with land parcel boundaries for day to day management using GIS techniques.

Master Plan of Indore City

Space Applications Centre (ISRO), Ahmedabad and Directorate of Town and Country Planning, Bhopal jointly took up the responsibility of preparing the "Master Plan of Indore city - 2011" using Remote Sensing and GIS techniques (Pathan et al., 2004). Entire Indore Planning area admeasuring 650 sq. km has been selected for the study. In the preparation of Master Plan of the city, one needs to assess the resources and their utilization with respect to the growing population in the area. Therefore, the population for the year 2011 was determined on the basis of standard statistical methods. It has been observed that the population of the Indore city would go from about 18 lakh (2001) to about 27 lakh by 2011. The additional area required for urban development to meet the needs of 9 lakh population over the year 2001 was determined based upon a population density of 100 persons per ha (Urban Development Planning, Formulation and Implementation [UDPFI] guidelines). As per the density adopted, the additional area required for urbanization is 8500 ha. To meet the additional demand of this land, an integrated analysis based upon different physical characteristics of the terrain viz. land use/land cover, hydrogeomorphology, groundwater prospects, natural hazards such as flood, erosion, earthquake etc. and environmental parameters such as air quality and water quality and infrastructure facilities was carried out employing a multi-variate index analysis approach (Pathan et al., 1989 and 1993). In all three planning scenarios have been generated on the basis of integrated analysis in GIS environment and were discussed with concerned authorities and finalized one scenario best suitable for implementation. The best suitable areas for urban development were later selected for the purpose of land use allocations with respect to the development of residential, industrial, public and semi-public, recreational, transformational etc., based on UDPFI norms. The 'Master Plan of Indore city' is presented in **Figure-2**. The entire plan was prepared in a record time of nine months with less cost and man power. The methodologies developed under this project are useful to prepare regional plan, master plan of any city in the country. The query shells and application shells developed under this project are extremely useful to retrieve the data on any administrative unit and help them in carrying out suitability analysis, environmental sensitivity analysis in an automated way. Hence, the case study of Indore has been considered as a model work to be adopted in the country.

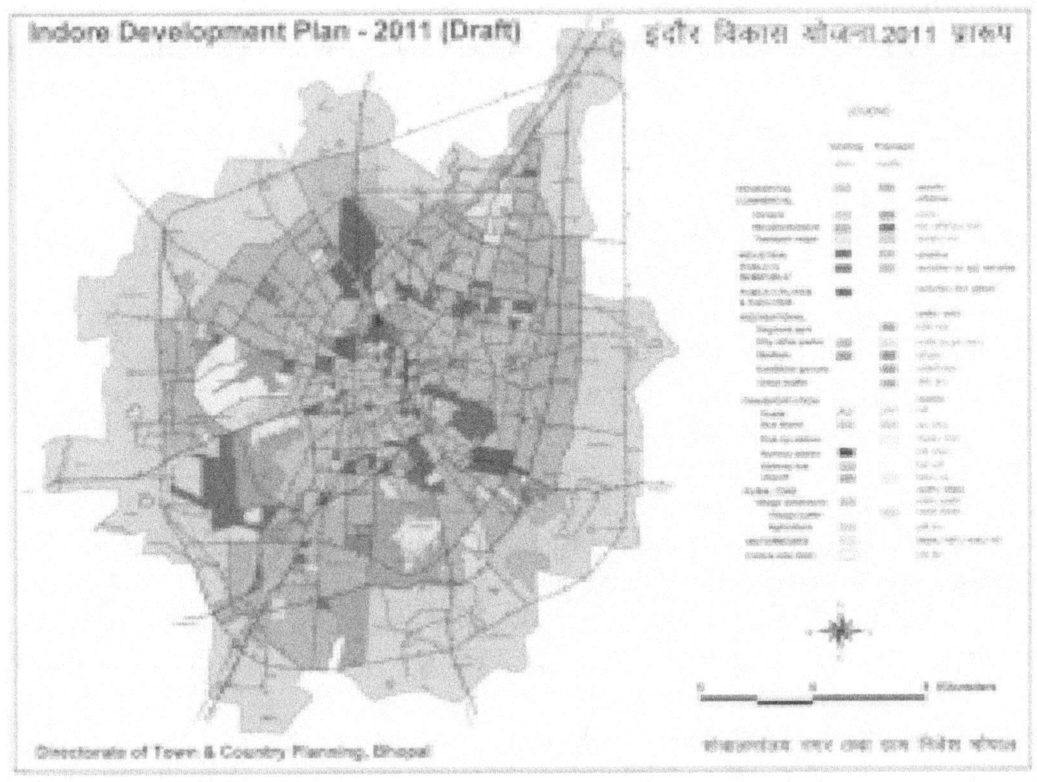

Figure-2 : Master Plan of Indore City

Infrastructure and utilities planning

Infrastructure and utilities viz. transportation, water, sewage, gas, electric power supplies and natural resources management about wells, reservoirs and water pollution basically associated on the optimum arrangement of networks. Thus a successful management of infrastructure development and utilities demand a concise yet comprehensive model of networks and the flow through them. Networks can store the information related to the linear features and the stored data can be utilised for a variety of planning purposes. For example, i) generation of optimum paths for emergency services thereby minimizing response time, ii) optimum alignment of utility lines (power, gas, water etc.,) and iii) service and parking demands for public facilities such as educational and research institutes, commercial centres, recreational areas etc. Some studies carried out demonstrating the potential of GIS Network techniques are given below.

Optimisation of Transport Service using GIS techniques

"GIS Network" tools have been employed for integration of spatial and aspatial data, facilitating retrieval of data with much greater speed, manipulation of data sets to arrive at alternate scenarios in various utilities planning including optimum transportation planning. In this regard, the work carried out at the Space Applications Centre not only demonstrates the capabilities of this technology but also its utility in some end-use applications related to sustainable transportation planning. The importance of network module in GIS has been presented with a specific study related to the optimisation of SAC transport service and GSRTC-ST transport service. An integrated analysis has been carried out to determine the optimum paths for Space Applications Centre (SAC) transport service and later the methodology has been tested on Gujarat State Road Transport Service (GSRTC) to validate it. The integrated procedures have helped to save travel time, distance, number of buses and cost. In case of SAC transport service, the number of buses has come down from 17 to 16 and a saving of Rs. 50,000 on oil consumption. Similarly GSTRC has saved Rs.1,50,000 per annum and number of buses have come down from 19 to 15. From these studies, it has been observed that i) the number of buses for the transport services could be reduced, ii) the travel distance can be brought down, iii) the travel time has also come down and iv) the diesel consumption per day has been considerably reduced. This enables GSRTC in particular, to save an amount of about Rs. 1,50,680 (Rupees one lakh fifty thousand six hundred and eighty only) per year on fuel consumption itself. This saving is variable as the diesel cost varies from time to time. Moreover this saving does not include the savings from wear and tear. Thus, the organisations can save much more amount not only on oil but also on wear and tear. The methodology developed in these studies is oriented towards planning, administering and operational management of resource facilities and can be applied to any service which is operated on the networks i.e. water supply, telephones, power supply, gas supply, sewerage etc. (Pathan et al, 1993, 2003).

Optimisation of Pipeline alignment using RS and GIS techniques

Another study has been carried out at SAC to align the pipeline in an optimum way using RS and GIS techniques. The approach for the optimum alignment has been followed on the basis of a multi-layer, multi-criteria analysis to delineate optimal routes for gas transmission pipeline from Myanmar to India connecting origin and destination points at Sittwe and Sasaram respectively. Two territorial options of routing through Myanmar-India and Myanmar-Bangladesh-India have been considered. Satellite imagery, existing maps and web resources were used for data base creation and analysis in GIS environment. Basic technique of optimization comes from shortest path algorithm on a cost surface generated by integration of all cost contributing layers. The desktop simulation study brings out that the routing through Bangladesh results in shorter length (1250 km) as compared to India–Myanmar routing (1544 km), the difference amounting to 24% increase in material cost i.e. pipe length. However the difference in relative construction cost is accentuated for the latter by more than 100% due to high and undulating terrains coupled with flood plains (Dubey et al., 2005).

Urban Information System

While a large amount of data is generated at various levels for urban planning and management it remains uncoordinated and often redundant to support decision making, as it is not available as a comprehensive compilation. Thus urban policymaking based on the linkages between various sectors e.g. health, education, physical infrastructure etc. leading to development of indices is not possible. On the other hand, planning activities involving spatial databases and decisions are not correlated with the sectoral/departmental data generated. In view of the developments in information technology in terms of HW, SW and networking it is now imperative that the country exploits the available technologies in order to enhance the efficiency in both planning and management of urban settlements.

In order to meet the above requirements, Ministry of Urban Development and Poverty Alleviation, Govt. of India has launched a programme called "National Urban Information System (NUIS). The objectives of this programme are :

- Develop attribute as well as spatial information base for various levels of urban planning.
- Use modern data sources such as Satellite and Aerial platforms.
- Integrate conventional data sources with modern data sources to develop GIS database.
- Develop standards with regard to database, methodology, equipment software, data exchange format etc.
- Develop urban indices to determine and monitor the health of the towns and cities
- Provide decision support system for planning at various levels.
- Build capacity among town planning professionals in the use of modern automated methods.
- Decentralize data generation, storage and manipulation at various levels of planning.

The data bases, the integration procedures and associated query shells and models developed under this programme will held in realizing the goals of various urban planning departments in the country.

Conclusions

From the above mentioned studies, it is very clear that the spatial information has been the most effective means of depicting events over space and time. Although, the significance of spatial information has been recognized in the past, with the advent of Geoinformatics techniques, the process of systematic spatial information acquisition has now become much easier. As Geoinformatics holistically comprise various frontier technologies such as Remote

Sensing, Geographic Information System (GIS), photogrammetry, Global Positioning System (GPS), etc., it aids to a large extent in various kinds of planning processes, decision-making and implementation of activities for the sustainable urban development planning and management in the country.

The availability of high-resolution satellite data of the order 1m and above has renewed interest in the development towards the preparation of Master Plans and Zonal Plans in the country. While full automated processes are still to come, significant levels of automation have been introduced in database generation and mapping to cater the needs of urban planners and administrators for their day to day exercises. The development of spatial databases under NUIS project employing RS and GIS techniques would aid as a decision-making tool and more in the context of assisting planning for various developmental activities in the urban areas.

Acknowledgements

I am extremely grateful to Dr. R.R. Navalgund, Director, SAC for his constant encouragement and keen interest shown in carrying out different urban studies at SAC. I express my sincere thanks to Dr. K.L. Majumdar, Deputy Director, RESIPA-SAC and Dr. Ajai, Group Director, FLPG for their guidance in successfully completing different urban projects. I also extend my sincere thanks to Shri R.J. Bhanderi, Scientist, RESIPA for his help in various stages of different urban studies.

References

Anon, 1997, Revised Development Plan of AUDA using RS and GIS techniques, p155

Dubey R.P., Maroo D.A., and Singh C.P., 2005, Remote sensing and GIS based pipeline alignment technique: A case study for India-Myanmar gas pipeline, Petrotech 2005 ID 337.

Pathan S.K. et al., 2004 'Urban Planning with specific reference to Indore Master', Proc. of GSDI-7 International Conference on "Spatial Data Infrastructure for Sustainable Development".

S.K. Pathan, "Potential of GISNetwork in Transportation Planning with a specific case study", Proc. of International Conference on "Transportation and Mobility : Strategic Initiatives using RS and GIS, 2003.

Pathan S.K., Shukla, V.K., Patel, R.G., Patel B.R. and Mehta, K.S., 1991, Urban Land use Mapping: A case study of Ahmedabad city and its environs, Photonirvachak, Journal in Indian Society of Remote Sensing, Vol. 19, No.2, 1991, pp: 95-112.

Pathan S.K., Navalgund R.R. and Pramod Kale, Potential of GIS Network for perspective planning -A case study", Proc. of Western Regional Convention 93, Computer Society of India, December 1993.

Pathan S.K., 1992, Remote Sensing in Urban land use mapping with data from different sensors, National Resources Management - A new perspective, NNRMS, pp371-376.

Pathan S.K., Jothimani P., Pendharkar S.P. and Sampat Kumar D., 1989, "Urban land use mapping and zoning of Bombay Metropolitan Region using Remote Sensing data, Journal of Indian Society of Remote Sensing, Vol. No. 17 (3), pp 11-22.

Dhinwa, P.S., Pathan S.K., Rao Mukund, Premnath, Chotani M.L. and Sinha R.L.P., 1994 "Land use change analysis of Bharatpur District using GIS techniques", Journal of Indian Society of Remote Sensing, Vol. No. 20, No. 4., pp 237-250.

Pathan et al, 1993, "Urban growth trend analysis using GIS techniques-A case study of the Bombay Metropolitan Region, International Journal of Remote Sensing, Vol.14, No.17, pp:3169-3179

Mukund Rao, Pathan S.K., Sastry S.V.C., Dhinwa P.S., Sampat Kumar D., Patkar V.N. and Phatak V.K., 1992, "A weighted index model for suitability assessment - A GIS approach, International News, ARC News, 14 (1) 28.

Pathan S.K., Jothimani P., Choudhary G.K., Som N.N and Mukherjee Kalyan, 1992, "Urban land use suitability analysis - A case study of Calcutta city and its environs", Journal of Indian Society of Remote Sensing, No. 20 (2&3), pp 73-84.

Population Growth And Land Transformation In Gurgaon, Haryana : Spatio – Temporal Analysis Using Remote Sensing And GIS Techniques

Rupesh Kumar Gupta

[+] Research Fellow, K.M.College, University of Delhi, Delhi-110 007

V. Raghavswamy

[*] Group Director, Land Use & Urban Studies, National Remote Sensing Agency, Balanagar, Hyderabad – 500 036, AP

ABSTRACT

In developing countries, rapid population growth is causing a decline in the per capita arable land transformation (man/land ratio) from rural to urban and also due to increasing migration of people to urban centres. Statistics suggest decreasing per capita land from 0.5 hectare to 0.2 acre by 1992. If the current trends in population growth continue, it is estimated that by 2050, the amount of arable land will be just over 0.1 hectare per person.

Gurgaon town is situated in the urban shadow zone of the Delhi. Its population has grown from 0.57 lakh in 1971 to 1.74 lakh in 2001. In addition, the pressure of continuously growing metropolitan city is also changing the urban morphology and urban landuse of the town and its neighbourhood. The present study is an effort to understand its population growth and the land transformation due to urbanization in space and in time.

Multi resolution and multi temporal satellite data of IRS(1B) LISS-II; IRS(1D) LISS-III, and PAN of 1993 to 2003, map data of 1971 and ground truth verification carried upto 2005 form the basic database. The spatial and temporal changes in the various land uses have revealed that, Gurgaon and its surrounding region have been growing rapidly especially during the last decade, and the transformation in land use patterns have been observed more along the major transport corridors, towards Delhi and Faridabad.

Multi resolution and multi temporal satellite data of IRS(1B) LISS-II; IRS(1D) LISS-III, and PAN of 1993 to 2003, with available map data of 1971 period and ground truth verification carried upto 2005 from the basic database.

The spatial and temporal changes in the various land uses have revealed that, Gurgaon and its surrounding region have been growing rapidly especially during the last decade, and the transformation in land use patterns have been observed along the major transport corridors towards Delhi and Faridabad.

Introduction

In the emerging scenario of increasing population size and urbanisation, around 2.9 billion people of the total global population lived in urban areas in the year 2000 A.D. and this is expected to rise to 5 billion by 2030. This amounts to 47% of the world population in urban areas, which is also expected to rise to 60% by 2030 (UN 2001). At the current rate of change, the world population is expected to be divided in equal halves between the rural and the urban by 2007.

Although the population living in urban areas is rising, the proportion of population living in urban agglomerations or metro cities is still small. In 2000, only 6.5% of the urban population of the world lived in cities with population size more than 5 million. This figure is likely to reach 8.4% by 2015. It is 52.5% (2000) in settlements with less than 0.5 million which is likely to be still around 50% in 2015 (Table-1).

One of the impacts of increasing urban population is on the cultivated land, which experiences a decline on account of industrial, residential, commercial, institutional activities and urban infrastructural activities. An estimated 5 million to 7 million hectares of farming land disappear world wide each year because of accelerating land degradation and rapid urbanization. As populous agricultural areas become even more crowded, arable land is likely to come under increasing pressure. India is no exception to this phenomenon due to increasing urbanization and market economy.

Table 1 Selected Indicators for the Urban and Rural Population by Development Group, 1950-2030.

Development Group	Population (billions)				Growth Rate (%) Doubling Time (Years)			
	1950	1975	2000	2030	1950-2000	2000-2030	1950-2000	2000-2030
A. Population size and growth								
Total Population								
World	2.52	4.07	6.06	8.27	1.75	1.04	40	67
More Developed Regions	0.81	1.05	1.19	1.22	0.76	0.07	91	998
Less Developed Regions	1.72	3.02	4.87	7.05	2.10	1.24	33	56
Urban Population								
World	0.75	1.54	2.86	4.98	2.68	1.85	26	38
More Developed Regions	0.45	0.73	0.90	1.00	1.40	0.38	50	185
Less Developed Regions	0.30	0.81	1.96	3.98	3.73	2.35	90	29
Rural Population								
World	1.77	2.52	3.19	3.29	1.18	0.10	64	714
More Developed Regions	0.37	0.31	0.29	0.21	-0.65	-1.09	—	—
Less Developed Regions	1.40	2.21	2.90	3.08	1.40	0.20	44	352

Development Group	Percentage Urban				Urbanisation Rate(%) Doubling Time (Years)			
	1950	1975	2000	2030	1950-2000	2000-2030	1950-2000	2000-2030
World	29.8	37.9	47.2	60.2	0.92	0.81	75	86
More Developed Regions	54.9	70.0	75.4	82.6	0.63	0.31	—	—
Less Developed Regions	17.8	26.8	40.4	56.4	1.63	1.11	42	62

Sources: United Nation Population Division, World Urbanization Prospects: The 2001 Revision.

Gurgaon, located around 35km. from the National Capital of Delhi has been experiencing rapid changes in population growth and land use during the recent years. Its accessability to international airport and good infrastructure has attracted several multinational companies to set up their industrial units in the Udyog Vihar-industrial area besides having other commercial activities. Geographically, it is located on a rolling plain surrounded by Aravali hills along the Western and Northern sides.

Objective

- To measure the extent of urbanization and urban growth in Gurgaon.

- To study the spatio-temporal land transformations due to urban sprawl.

Database

- IRS (1B) LISS-II, IRS (1D) LISS-III and PAN data of 1993 and 2003; map sheets of 1971, revenue (village) maps; land records data and census data supported by ground/field data.

Research Design and Methodology

The work has been based carried using satellite data and map sheets and village maps. The data has been processed using Geomatica, Erdas and Arc-GIS software. GIS helps to display the spatial location and extent of urban sprawl and the different categories of the land use at different periods of time, which is essential to understand the trend and rate of the change. The integration of remote sensing and GIS also strengthens the decision support process on a scientific basis for carrying urban planning and urban management.

The data collection involved primary data and secondary data. To understand the complexity of dynamics of land use changes expansion pattern of the city and land transformation, a few indicators were examined. The indicators such as land use, roads, railway network and the agricultural area were captured from map sheets and imagery and each of the layers were digitized. The extension of agriculture land during the last three decades (1971-2002) is determined by computing the area from the digitized map sheets, imagery and compared it with the areas of different time periods. The land use classification is shown in Table-2 and the description of the broad land use class in Table-3.

Table 2 Scheme of Urban Land Use Classification

LEVEL I	LEVEL II	LEVEL III
* Urban or Built-up	* Residential area	* Dense
		* Moderate
		* Slums
		* DLF layout
		* Residential layouts/ open space
	* Commercial	
	* Industrial	
	* Transportation	* Railways
		* Main roads
		* National highway
		* Others roads
	* Institutional	
	* Recreational	* Parks/Gardens
		* Playgrounds
		* Stadium
		* Golf course
* Services	* Railway station	
	* Police station	
	* Post office	
	* Telegraph office	
	* Hospital	
	* Wireless station	
* Agricultural land	* Cropland	
	* Fallow land	
	* Plantation	
* Wastelands	* Scrub land	
* Forest	* Water logged	
	* Hills/Barren rock	
	* Trees, Green cover	
* Water bodies	* River/Streams	
* Rural	* Tanks/ponds	
	* Rural settlements	

Table 3 Land use classes considered in image classification

Land use class	General description/Subclass
* Built-up Land	Include Dense, Moderate, Sparse as well as Rural settlements, Industrial, Institutional, Commercial, Recreational, Transportation and utilities.
* Agricultural Land	Cropland, Fallow land and Plantation
* Forest	Trees, green cover
* Wasteland	Scrub land, Water logged, hills / barren rock
* Water Bodies	River, tank/lakes

Population growth in Gurgaon

Gurgaon has been a small but important urban centre in the neighborhood of Delhi. A district headquarter; its importance lay in providing space for the defence and wireless station during the British period. The airport later turned to be an important international airport for Delhi.

Table-4 shows that the city had a small population of 4765 in 1901. The population grew at the rate of 1.46% during 1901-1911. It faced a decline during 1911-1921 (-6.48%) in line with the rest of India, a period when epidemics, took a heavy toll of India's population. In the subsequent decades of 1931 and 1941, its rate of growth had been higher than national average.

Table 4 Urbanization Trends in Gurgaon 1901 – 2001

Year	Urban population	Decadal Growth%
1901	4765	
1911	5461	14.61
1921	5107	-6.48
1931	7208	41.14
1941	9935	37.83
1951	18613	87.35
1961	37868	103.45
1971	57151	50.92
1981	89115	55.93
1991	121486	36.32
2001	173542	42.85

Sources: Census of India, Economic Division, Govt. of India, 2001

Subsequent migration of people after partition resulted in a sharp increase in population, reflected in the next two decades (Table 3). Population grew by 87.35 % between 1941-51 and by 103.45% during 1951-61. The location of Gurgaon vis-à-vis Delhi has always been a positive factor in attracting people and industrial activity, resulting in higher growth rate than the national average. Between 1981-91, a lower growth rate of 36.32% has been recorded.

A change in economic policies of the 1990's leading to a trend in globalization, led to the influx of multinational companies and increase in residential; commercial; industrial space and further increase in population. The growth rate was 42.85% between 1991-2001 and the population of Gurgaon city in 2001 census is enumerated to be around 1.74 lakhs.

The villages in Gurgaon region have shown (Table-5) a much higher growth rate than Gurgaon city during 1981-91 - the average being 64.71% as compared to 36.32% for the

Table 5 Population Growth in the Villages of Gurgaon Region

Villages	Total Population			1971-81	1961-91	1971-81	1981-91
	1971	1981	1991	Decadal Increase	Decadal Increase	Growth (%) Rate	Growth (%) Rate
Begumpur Khhlola	889	1044	1416	155	372	17.40	35.63
Chakkarpur	1509	1806	2525	299	717	19.81	39.65
Fazilpur Jharsa	836	1150	1532	312	382	37.23	33.2
Gurgaon Rural	8080	32956	82710	24876	49754	307.00	151.00
Islampur	1182	1632	2436	450	804	38	49.26
Jharsa	7020	7506	8480	486	974	6.9	12.97
Kadipur	787	2646	3310	1859	664	236.0	25.00
Naharpur Rupa	507	622	1586	115	964	22.68	154.9
Nathupur	2133	3076	3508	943	432	44.21	14.00
Salokhara	714	1065	1986	351	921	49.15	86.47
Shamashpur	593	752	946	159	194	26.8	25.79
Sikandarpur Ghosi	997	1679	2772	682	1093	68.4	65.10
Sirhaul	1947	2098	2638	151	540	7.75	25.74
Tigra	712	1031	1333	319	302	44.8	29.29
Tikri	492	658	701	166	43	33.74	6.5
Wazirabad	3871	4754	5670	883	916	22.81	19.27
Total	32271	64477	123549	32206	59072	99.7	91.62

Sources: District Census Handbook (1971,1981,1991), Gurgaon District, pub. by Govt. of Haryana, India.

city during 1971-81. Further during 1981-91 four villages have shown a growth rate of more than 70%. During 1971-81, population of Gurgaon city grew by 56%, while the population in the villages grew by 35.3%. The estimates indicate that the rate of growth of population in the villages is likely to be higher still during 1991-2001 on account of industrial, residential and commercial activities created in the city region as a result of liberalization of economy.

Land Transformation : Spatial and Temporal.

Considerable changes in land use has occurred during the last few decades. The change may be in a specific area or the entire region. The change from rural to urban land use is so fast that the resultant need and complex uses coupled with shortage of land have led to speculation and increase in land values. The ever-growing difference between the demand and supply of house sites has increased the cost of land in the city which has ultimately led to pressure on fringe areas which has given rise to development of residential areas and industries. The Table-6 show the urbanization trend in Gurgaon during the last one century (1901-2001).

Table 6 Land transformation in different area (in percent)

Code	Land use categories	1971-93	1993-2002
1	Built up Land	11.36	47.82
1-5	Built up Land -Others	-	0.42
2-1	Agroland- Built up Land	39.42	55.64
2	Agroland	63.35	5.84
2-3	Agroland-Water	-	0.11
2-4	Agroland-Wasteland	-	0.24
2-5	Agroland-Others	1.78	1.43
3-1	Water bodies- Built up Land	-	0.15
3	Water bodies	-	0.014
4-1	Wasteland- Built up Land	-	2.45
4-2	Wasteland-Agroland	-	1.61
4	Wasteland	6.75	4.77
4-5	Wasteland-Others	-	0.96
5-1	Others- Built up Land	0.35	3.21
5-2	Others-Agroland	0.89	1.07
5-4	Others-Wasteland	0.18	-
5	Others	2.62	1.04
	Total	126.77	126.77

Sources: Map sheets 1971, IRS (1B) 1993,IRS (1D) 2002 & IRS 1D PAN Merged data.

Table-6 reveals that three subcategories of land-use have undergone substantial changes. The maximum change has occurred in the agriculture land, which has declined, from 63.35 sq.km. of the total land in 1971 to 5.84 sq.km. in 2002. Next in order is the 'Built-up land', which has increased from 11.36 sq.km. to 47.82 sq.km. during the same period. The next significant change occurred is the conversion of agricultural land to built up land, from 39.42% to 55.64%.

The statistics reveal that out of the 126.77 sq.kms. as total area of Gurgaon city/region, nearly 81% was under agriculture in 1971. It has reduced to (50.67%) in 1993 and to 26.5% in 2002. The rate of decline has been higher during the last decade. Between 1971-93 around 38.43 sq.kms. of agricultural land was lost and between 1993-2002 around 30.65 sq.kms. - Table-7 & 8.

Table 7 Comparison of areas under different land uses

Land use class	1971		1993		2002	
	Area (sqkm)	%	Area (sqkm)	%	Area (sqkm)	%
Built-up land	11.36	8.96	51.14	40.34	84.2	66.42
Agricultural land	102.67	80.99	64.24	50.67	33.59	26.50
Waste land	7.48	5.90	6.97	5.50	5.01	3.95
Others	5.26	4.15	4.42	3.49	3.97	3.13
Total	126.77	100	126.77	100	126.77	100

Sources: Map sheet 1971, IRS(IB) 1993, IRS(1D) 2002 & IRS 1D PAN Merged data of 2002

Table 8 Comparison of changes areas under different land uses

Land use class	Changes	Area	Percentage of Changes	
	1971-93	93-2002	1971-93	1993-2002
	Area in (sqkm.)	Area in (sqkm.)	%	%
Built-up land	39.78	33.06	50	50
Agricultural land	38.43	30.65	48.30	46.35
Waste land	0.51	1.96	0.65	2.96
Others	0.84	0.45	1.05	0.69
Total	79.56	66.12	100	100

Sources: Map sheets 1971, IRS(1B) 1993, IRS(1D) 2002 & IRS(1D) PAN Merged data of 2002.

The area under 'built up land' category increased substantially from 11.36 sq.kms. (8.96%) in 1971 to 84.2 sq.kms. (66.42%) in 2002. The rate of increase of built up area has been higher between 1993-2002 (3.6 sq.kms. per annum) by an addition of 33.06 sq.kms. compared to 1971-93 period (1.81 sq.kms. per annum) during which 39.78 sq.kms. was added. There was 33% decline in wasteland (from 7.48 sq.kms. to 5.01 sq.kms.), more rapidly during the last decade than in the first two decades. 'Other land uses' have reduced by 25% from 1971 to 2002 (from 5.26 sq.kms. to 3.97 sq.kms.) with a rate of decline from 10% during 1993-2002, as compared to 15% during 1971-1993.

Figure-1 to 3, depict land use pattern in and around of Gurgaon in 1971, 1993 and 2002 and Figure-4 shows the land use changes during the last decade 9193-2002) which is still continuing. This can be attributed to rapid outsourcing of economic activities and increase in the demand for residential and commercial purpose. The villages in Gurgaon region can broadly be classified into three categories based on the process of their transformation : Rural – dominated with agricultural land and primary activities; transitional semi rural to semi urban – dominated with built up land and territory activities. The process of urban sprawl is clearly brought out in Figure-1 to 3.

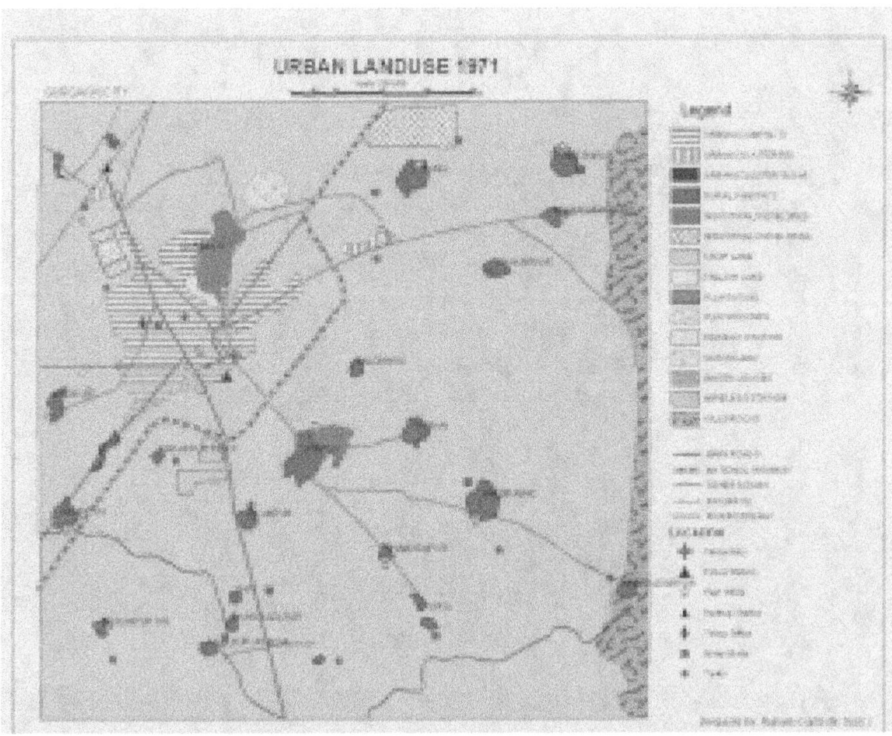

Fig. 1 Land Use Map1971

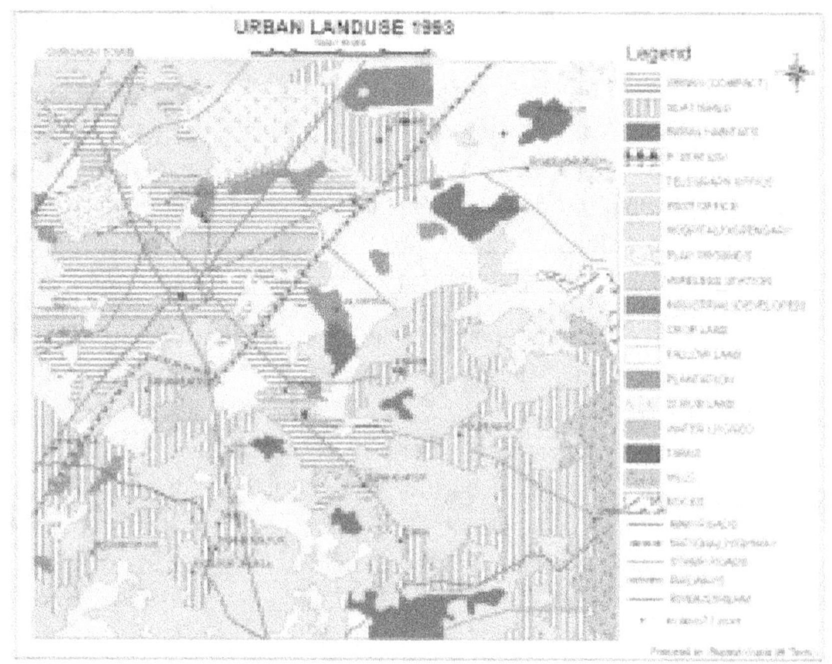

Fig. 2 Land Use Map1993

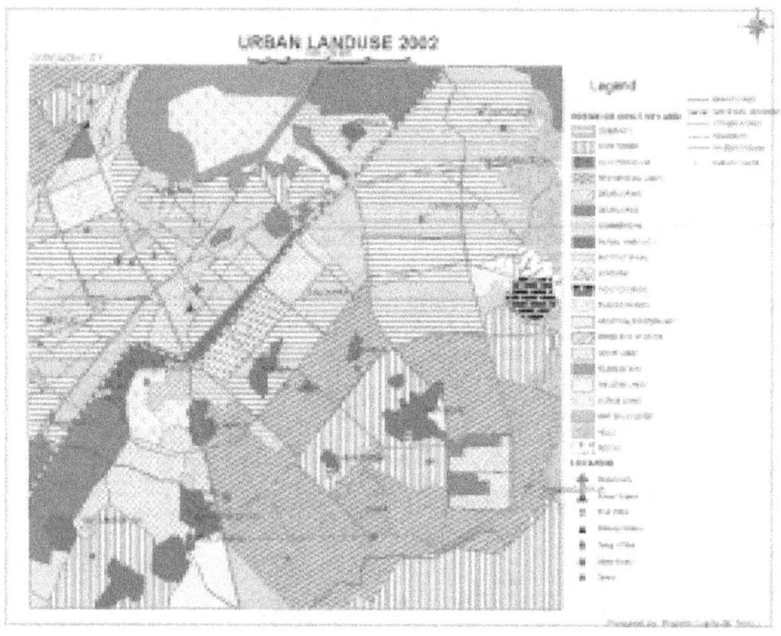

Fig. 3 Land Use Map 2002

Figure-4 (urban land use change 1993-2002) highlight the following :

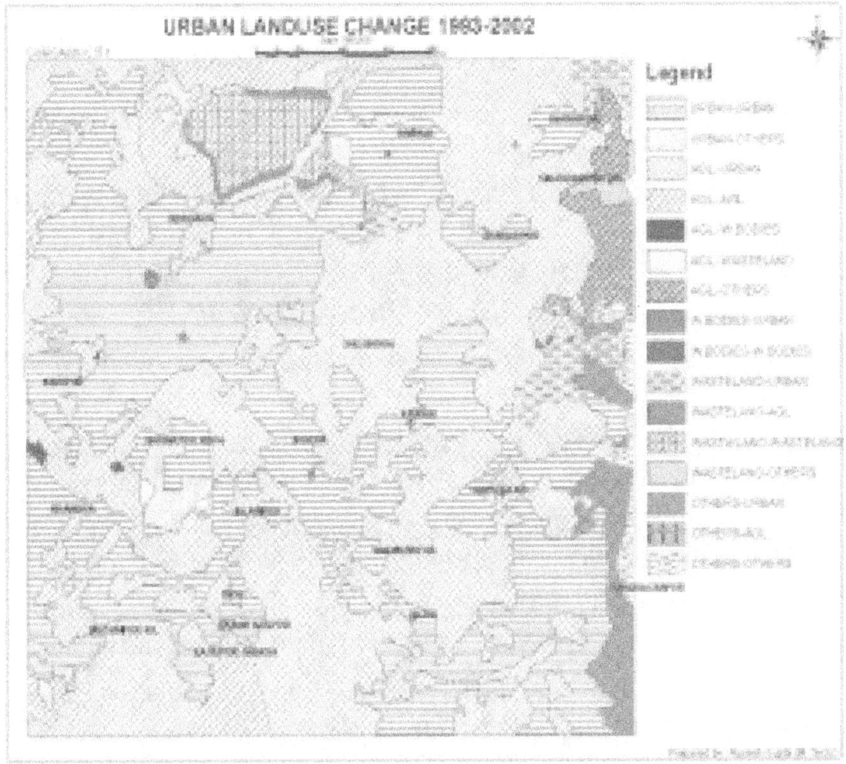

Fig. 4 Land Use Change Map 1993-2002.

- Major conversion of agricultural land into commercial, residential and the other urban land uses.

- The impact of mega city Delhi and development of physical infrastructure especially the transport system has triggered the land use changes.

- The open areas, greenery of surrounding area as well as its vicinity to Delhi are some factors, which attract the people towards this satellite city.

Acknowledgement

Thanks are due to Prof.S.K.Mukherjee Vice Chancellor of BIT, Mesra, Ranchi and Director NASA, Hyd for giving the opportunity to carry out this work and also to the Dept. of Science and Technology, Govt. of Haryana, for providing the data. Thanks are also due to Dr.M.S.Nathawat, BIT, Mesra and to Dr. Jitendra Prasad, HARSAC, Hisar for their guidance and cooperation.

References

1. Amarsaikhan.D&Ganzorig, M (2002) Urban Change Study Using remote sensing And GIS, ACRS, Institute of Informatics and Mongolian Academy of Sciencesav.Enkhtaivan-54B, Ulaanbaatar-51 MONGOLIA.

2. *ACRS, (2002),* Remote Sensing and GIS for Sustainable Development in Haryana State, India, *HARSAC, Hissar.*

3. *Clark, KC.&Silva, EA (2001),* Calibration of the SLEUTH urban Growth Model for Lisbon   and Porto, *Portgal, Univ. of California/Massachusetts.*

4. *Gautam, NC (2002),* Methodology for land use planning - A systematic approach, *CLUMA Publication, Hyderabad, India.*

5. Govt. of India. (2002-2003), Ministry of Finance, Economic Survey, Economic Division, New Delhi.

6. Govt. of Haryana (1973) District Census Handbook 1971, Gurgaon District, Part X-A&B, Series-6, pp.156-171, pub. by Govt. of Haryana.

7. Govt. of Haryana (1983) District Census Handbook 1981, Gurgaon District, Part XIII-A&B, Series-6, pp.44-49, pub. by Govt. of Haryana, India.

8. Govt. of Haryana (1993) District Census Handbook 1991, Gurgaon District, Part XII-A&B, Series-VIII, pp.108-117, pub. by Govt. of Haryana, India.

9. *Jensen, JR, (1996),* Introductory Digital Image Processing-A Remote Sensing Perspective, *Prentice Hall, New Jersey*

10. *Lay, JG (ACRS-2000).* A Land Use Change Study Using Cellular Automata, *Dept. of Geography, National Taiwan University.*

11. Raghavswamy V. et.al (2003). Very high resolution satellite data Urban applications; Geospatial Today, Vol.2, Issue-3, pp.25-27.

12. Saxena, A, Remote Sensing & GIS in Assessing Physical Transformation of Bhopal City. email:Arunasaxena2000@yahoo.com

13. Sudhira H.S. et.al. Urban sprawl pattern recognition and modeling using GIS, Centre for Ecological Sciences, IISc, Bangalore.

14. Tiwari D. P. ACRS(2002), Remote Sensing and G.I.S. for efficient Urban Planning. email: tiwari_dp@hotmail.com

15. United Nations, 'World Urbanization Prospects-The 2001 Revision' (UN Secretariat population Division), Data Tables and Highlights 2002. http://www.un.org/esa/population

16. *Yeh, A.G.O, and Xia Li, (2001),* "Measurement and Monitoring of Urban Sprawl in a Rapidly Growing Region Using Entropy", *Photogrammetric Engineering and Remote Sensing, vol.67 (1): pp 83.*

Impact of Pollution and Quality of Life of the Residents of Hyderabad

Afzal Sharieff,

Associate Professor, O.U. Hyderabad

Mohammed Akhter Ali,

Dept of Geography, O.U., Hyderabad

ABSTRACT

Scattered location of industries particularly hazardous chemical industries in Hyderabad region in Telangana of Andhra Pradesh, are primary source of pollution and cause of contamination of soil and water in the region. The source and pattern of land and water bodies' contamination was traced and extent of affected area was quantified. Balanagar - Jeedimetla – Kukatpalli IDA which houses 41 polluting units, drain into Hussain sagar which is highly polluted as indicated by the reflectance anomaly between water lakes and polluted tanks.

Pollution of water bodies and disposal of industrial waste on ground has affected the groundwater quality in the downstream residential areas of Indranagar, Ashoknagar, Musheerabad, Himayathnagar, Chikkadpally and Narayanguda. In Jeedimetla groundwater is highly colored and unpotable. Unless concrete measures are taken to decontaminate the Kukatpally and Balanagar Nalas (drains) and Hussain Sagar Lake is restored, the Musi River would remain a receptacle of polluted water as highlighted in the news papers recently. There are many laws which regulate these polluting industries check, but lack of enforcement agencies, corruption, coupled with many other reasons are to be examined and reorient to enable environment free surrounding.

Introduction

Hyderabad occupies a unique position in India. It is one of the leading mega city of India. The primate city of Andhra Pradesh and strategically located software and industrial center in south India. Hyderabad with population of 5.6 million (2001) and a history of more than four hundred years representing socio-economic and ethnic character of both North and South India deserves to be called a city of 'cultural saga'. Hyderabad is located in the heart of Deccan trap land at a fairly high altitude of 540 m above Mean Sea Level (msl) and has a salubrious climate. It is located at the intersection of 17° 17' 30", 17° 27' 30" N latitude and 78° 22' 30" – 78° 37' 30" E longitudes, covering an area of 217 sq.kms. The general slope of the city is eastward- which can be observed by the flow of the river Musi form west to east. The Hyderabad city has undergone considerable political and socio-economic changes in the past. Today also0, it is under going rapid and dynamic change due to accelerated phase of industrialization, software development and tourism.

Hyderabad city is situated on the Deccan Plateau region in the Krishna river basin area. River Musi is a tributary of river Krishna that physically divides Hyderabad city into Hyderabad North- the new city, Hyderabad south -the old city or the historical core of the city. The catchment area of River Musi can be divided into two parts, namely South bank and North bank. The North bank covers about 60 percent of the incorporated area. The North bank is divided into Musi river drainage area and Hussainsagar drainage area.

Hussainsagar drainage area drains part of Hyderabad North and Secunderabad. It is about 50 feet higher than River Musi drainage area. The western portion is characterized by boulders and hills like Banjara and Jubilee hills. The general slope is from west to east.

Hussain sagar is situated between the twin cities of Hyderabad and Secunderabad. It was excavated in 1562 to store drinking water brought from the river Musi by Balakpur canal. However, with passage of time the lake lost its importance as a source of potable water. Today it is extensively used for washing, bathing and recreation. Total degradation of the lake is essentially a direct result of unplanned industrialization and consequent urbanization. The lake is 510 meters above the msl. Water shed of Hussainsagar is divided into four sub basin; Kukadpally, Dullapally, Bowanpally and Yousufguda. The highest peak in catchment is at 642 meters north of Nizampet and the lowest contour near tank bund at 500 meters. The effective north south drop thus comes to 142 meters covering a distance of 17 Kms. The lake is fed by four streams (Nalas); Kudakpally- 70 liters per day; Picket- 4 million liters per day; Banjara -6 million liters per day and Bulakpur 13.3 million liters per day. Of the four, Kukadpally is the main feeding channel that brings in major bulk of water into the lake. Today the water entering into the lake through Kukadpally has 55 MLD of domestic sewage and 16 MLD industrtial effluents. The Kukadpally stream has total catchment area of 168 sq.kms. The entire catchment has two major industrial areas; Kukadpally and Balanagar with major industries like IDPL, HAL, Alkali Chemical Corporation limited, Asbestos from public sector and number of those belonging to the private sector.

Initially the water body did absorb the pollution impact but its natural caring capacity was reached its limits, adverse effects of pollution started, in the form of deterioration of water quality, fowl smell, wild growth of macrophysics and breading vector like mosquitoes. Up to 1992 the pollution reached such a point that the lake becomes a cesspool of polluted water. The problem of cnvironmental degradation of lake was further complicated by two ever expanding of slums along western bank of the lake. One of these is opposite to the Rajbahavan and the other near Begampet bridge area at the rear shore of the lake. Ganesh idols immersion and washing activities along eastern waterfront are also the source of polluting of water body.

About 45 years back the lake was pollution free with good water quality and biodiversity. During last four decades there is progressive and today following classical symptoms of pollution are clearly visible. All these typical problems responsible for environmental degradation of a lake in urban environment were responsible for the slow destruction of the Hussainsagar Lake.

Vast Industrialization and immense urban growth have let thousands of tones of pollutants into the river causing a huge water and air contamination to the city. Expansion of the city is further causing deforestation, soil erosion and related problems and degradation of environmental quality. On the other hand unwanted settlements encroached in to the riverbed have made the course channels narrow. As the demand for housing is increasing day-by-day many people are dwelling in the affinities of stinking channels, knowing well it is not hygienic to live around. Streams and gutters act as storm drains often creating higher flood peaks in the local stream. August 2000 flood havoc is the latest example of devastation. Besides sewerage, industrial effluents from Jeedimetla and Bala Nagar, Kukadpally have been directed in the Hussain sagar and hence it is highly polluted. None of its channels in the twin cities are holding fresh water today and all the channels have turned to be sewerage canals carrying domestic sewerage and industrial effluents. Scattered location of industries particularly hazardous chemical industries in Hyderabad region in Telangana of Andhra Pradesh, are primary source of pollution and cause of contamination of soil and water in the region. The source and pattern of land and water bodies' contamination was traced and extent of affected area was quantified. Balanagar - Jeedimetla – Kukadpalli IDA which houses 41 polluting units, drain into Hussain sagar which is highly polluted as indicated by the reflectance anomaly between water lakes and polluted tanks.

Since the fore most factors responsible for all the degradation is haphazard siting of industries, the EIA is specially focused on siting of industries. The current picture of siting of industries and loopholes can be seen clearly. Siting of industries is still based on all economic parameters such as, nearest market, availability of cheap labor, good road network and power supply etc. Site clearance procedure has several loopholes, which cannot predict the environmental impacts. Even if the impacts are found during pre feasible EIA, there are many flexibilities which will subside the environmental priorities.

The entrepreneurs usually select a site for an industry and go for clearance. The clearing authorities will not have enough of scientific database to assess the environmental impacts and make a fast decision-making.

The land use land cover is the most important parameter for siting of industries, as the land use it self is sensitive to industrial pollution and sensitive areas should not be industrialized. In order to identify the sensitive areas to be protected, the land use land cover map is prepared. Agriculture is the predominant land use practice. Kharif crop is more practiced than double crop as most of the land is rain fed. Tank irrigation is also predominant in the city as the rainwater is harvested in tanks and used for agriculture. The land is mostly covered by scrubs followed by Barren /rocky/stony waste/sheet rocks. Forest occupies next place. Ever green forests are not observed in the district and all forests are of deciduous type usually open, dominated by forest scrubs.

The Sanat Nagar Industrial Estate : It was established in 1957 under cooperative sector. Area at that time was 0.3682 Sq kms. Total investment from the govt was 81.8 lakhs and by entrepreneurs were 1.5 crores. Main products were bicycles and their parts, transistors, radios, saw and razor blades, steel and wooden furniture, insulated copper, conduit pipes, copper and PVC wire, tin containers, medicinal products, agricultural equipment, sheet metal products. In 1962 another 0.72 Sq. Km. of area was acquired. The Hyderabad city was 10 km away from this Industrial estate in1957. By 1980, this became a part of city because of the rapid urban development. Around 1990 air pollution was reported by NEERI in this area. Again in 1995 from the quality reports of NEERI, it was observed the air and water qualities were deteriorating in the light of increased chemical industrial activity. The report of the ground water board of 1998 stated that the ground water in this area was highly contaminated and was not suitable for any of the purposes.

The Kukadpalli Industrial Estate : It was established in 1981, which was originally an agricultural area by then. Rapid land use changes have occurred. Agricultural land has been diminished since then. Ground water quality has deteriorated to the extent that the water has turned into rust color with objectionable odor, which cannot be touched even. This is because of the contamination from solid and wastewater dumps.

Pollution of water bodies and disposal of industrial waste on ground has affected the groundwater quality in the downstream residential areas of Indranagar, Ashoknagar, Musheerabad, Himayathnagar, Chikkadpally and Narayanguda. In Jeedimetla groundwater is highly colored and unpotable. Unless concrete measures are taken to decontaminate the Kukatpally and Balanagar Nalas (drains) and Hussain Sagar Lake is restored, the Musi River would remain a receptacle of polluted water as highlighted in the news papers recently. There are many laws which regulate these polluting industries check, but lack of enforcement agencies, corruption, coupled with many other reasons are to be examined and reorient to enable environment free surrounding. EPTRI and APPCB have to play an important role in combating this pollution threat to the residences of down stream of Hussainsagar. All the Industries letting industrial effluent into the Hussainsagar without treatment should be heavily penalized and further their units should be shut down. Municipal corporation should take a lead in evacuating the nalas and downstream of the Hussain sagar. One most important step which government can take up in combating this pollution threat would be either made the industrialists to stay in the industrial area itself or at eh downstream, which is getting polluted with the untreated effluent being released in the water bodies.

References :

Alam, S.M. 1985, Vulnerability and Resilience of Cities, the case of Hyderabad, India, UNESCO Sponsored project, p.40.

Ahsan, Mohammad. Ecology of Fresh water zooplankton (Hussainsagar and Saroonagar Lake) PhD thesis, Osmania University, Hyderabad.

Associated Industrial Consultants Private Limited, Hussainsagar Lake, Hyderabad. The profile of a highly entropic lake, ecology. Vol.7, No. 10.

Deccan Chronicle, Hyderabad, Rain could sweep industrial waste into water reservoir, Dated 8-6-2001.

Editorial Report, Environment and Urbanization – Volume 11, No. 2, October 1999.

Kadarkar, M.S. Muley, S.V. and Vasant Rao, Kukadpally Hussainsagar, Ecological studies on an industrially polluted stream and its impact on a freshwater lake in Hyderabad, Indian Association of Aquatic and Biologist, Hyderabad, Pub. No. 1.

Blue Print of Problematic Areas In Bangalore Rural District Identified Through Remote Sensing

Nagesh

[1] Consultant, Karnataka State Remote Sensing Applications Centre (KSRSAC), Bangalore

M Padmavathi, V Shreedhara

[2 & 3] Scientists, KSRSAC, Bangalore

H Honne Gowda

[4] Director, KSRSAC, Bangalore

ABSTRACT

About 63% of the geographical area (71% of net sown area) in Karnataka is in dry zones. On an average of 1995-96 and 1996-97 production, out of the total of 8.18 Mt of cereals, 62% and almost all the pulses of 1.65 Mt came from rainfed areas. Land use/land cover mapping using Remote Sensing techniques give information about existing land use/land cover, which has to be assessed further in terms of the seasons of the crops and sources of the irrigation to understand the land potential before suggesting the alternate land use practices. The objective of the study was to identify the rainfed and irrigated areas, sources of irrigation, wastelands and forest lands of Bangalore Rural district of Karnataka ($77^0 2'$ to $77^0 58'$ longitude and $13^0 29'$ to $12^0 14'$ latitude) using satellite data.

The results indicated that out of the total geographical area of 5,81,937 ha of Bangalore Rural district, area under agriculture was 3,22,464 ha (55.41%). Out of the total cropped area of 3,22,464 ha, 2,25,600 ha (69.96%) area was cultivated in kharif season. Out of the total kharif area of 2,25,000 ha, 2,21,740 ha (98.29%) was under rainfed cultivation and the remaining 1.71% (3860 ha) was under various irrigation sources such as canal, lift, well and tank. Out of the total cropped area of 3,22,464 ha, 96739 ha (30%) was under double crop under various sources of irrigation. The two major sources were tanks (50,204 ha, 51.90%) and wells (45,158 ha, 46.68%).

The information about the sources of irrigation is a crucial factor for the agricultural output. The sustenance of the well and tank irrigation depends on the rainfall as they are not assured sources unlike that under canal irrigation from reservoirs. Areas irrigated by wells and tanks and areas purely cultivated under rainfed conditions became problematic and become a cause of concern once the onset of the monsoon delays or rainfall is below the normal.

Introduction

One of the main challenges facing India in the coming decades is providing adequate food, economic and health security for millions of people in the country on continuing and sustainable basis. In spite of the Green Revolution, India's agricultural productivity is one of the lowest at 1.6t/ha compared to the world's average of 2.6t/ha and that of developed nations over 5t/ha. While the Green Revolution in the past enabled the country to increase its annual grain production from 55 Mt to 180 Mt, this very high technology practice, involving a large-scale use of chemical fertilizers and pesticides, high-yield cultivars and extensive irrigation combined with poor drainage characteristics, has turned vast tracts of once fertile agricultural land into saline and alkaline deserts.

Estimates indicate that in India, out of total of 160 M ha, 100 M ha of arable land has already been degraded, 50 M ha of which has been severely degraded and became unproductive. Reclamation of cultivable wasteland, which has been clearly identified by the remote sensing satellites, can at best add 20 M ha to the total arable land of 160 M ha. Even assuming that we exploit the total irrigation potential, by fully developing the water resources to irrigate 80 M ha of agricultural land as against the present 46 M ha, our total grain production per year can at best be increased to 250 Mt with the present agricultural practices.

About 63% of the geographical area (71% of net sown area) in Karnataka is in dry zones. With only about 23% of the area under irrigation, a substantial contribution to agricultural production comes from drylands. On an average of 1995-96 and 1996-97 production, out of the total of 8.18 M t of cereals, 62% came from rainfed areas whereas almost all the pulses of 1.65 M t was under rainfed conditions. This indicates the importance of dryland cultivation and the need to improve the irrigation capability.

Preparation of Agricultural Resources Inventory

Space remote sensing which can now provide synoptic and repetitive high-resolution imageries over large areas, has become a powerful tool for mapping spatial as well as temporal changes in soil characteristics, soil moisture and land-use patterns at micro-levels, to identify forestry, plantations, pastures, single and double cropped areas, cultivable wasteland and fallow residual land. Remote sensing satellites have proved their immense potential for mapping surface and under ground water resource and for predicting the acreage and yield of various crops in advance.

The technology of predicting, monitoring and managing drought using remote sensing imageries has greatly assisted planners in evolving strategies to deal with such extensive disasters and mitigate their effects. Land use/land cover mapping using Remote Sensing techniques is conducted in the state by KSRSAC which gives information about information on existing land use/land cover and pattern of their spatial distribution which forms the basis for any developmental planning. The current land use has to be assessed further in terms of the seasons of the crops and sources of the irrigation to understand the land potential before suggesting the alternate land use practices.

Utilization of the maps

The seasonal details of the crops, sources of irrigation and information about plantation crops are obtainable directly from this data and this can serve as base line data map. The study areas, if spread over a large number of districts/water sheds, will cover a variety of agro-climatic and terrain conditions. The data could be compiled to get a zonal picture, which would be useful for monitoring the implementation of zonal perspective plan over years. Since Remote Sensing Techniques will be used to assess the resource status it is appropriate that same means should be used for monitoring also. The feedback shall help in estimating benefits of plans and also help in making modifications in methodology for future studies.

The maps can help in planning at hobli/district level and if the present land use and irrigation status is considered sub optimal, a few possible options can be worked with an aim to achieve optimality within the frame work of sustainability of production. The threshold limit of a particular parameter vis-à-vis its consideration for a particular land use practice varies from area to area. While making these recommendations land form, soil, slope, ground water potential and quality, rainfall and agro-climatic zones, present land use etc., are considered.

While making alternate recommendations for land use, futuristic considerations such as exploitation of ground water, if presently not exploited, and possibility of adopting more efficient system of irrigation and water management and other site improvements through soil and water conservation can be planned using this information.

Finally achievement of the above said objective would aid the planners, decision makers and implementation authorities to:

(a)   Coordinate, formulate and implement the land management policies and prepare a plan of action at taluk level covering a time frame ;

(b)   Monitor the action taken thereon in a fruitful manner and ;

(c)   Create a greater awareness of the advantages of identification of problematic areas at all levels through satellite remote sensing.

Objective : To identify the rainfed and irrigated areas, sources of irrigation, wastelands and forest lands of each taluk of Bangalore Rural district using satellite data.

Study area : The study area is Bangalore Rural district which lies in $77^0 2'$ to $77^0 58'$ longitude and $13^0 29'$ to $12^0 14'$ latitude.

Data used :

(a) Satellite data: Satellite data for 2000-01 (IRS 1C/1D PAN + LISS III merged data and LIIS III) generated for the land use / land cover thematic map under SNRIS project would be used.

(b) Ancillary data : Major and medium irrigation command boundaries, minor irrigation command boundaries and well data from Dept. of Mines and Geology.

(c) Ground data: The ground truth data was collected for the study area. The sample sites were selected in such a way that they are easily identifiable and aid as ground control points for identification on satellite imagery.

Methodology

On screen delineation of the proposed classification was carried out using ERDAS Imagine software. The details of classification are as follows:

Rainfed land : From land use/ land cover data rainfed agricultural area were identified by referring two season data.

Irrigated land : From land use/ land cover data irrigated agricultural area were identified by referring two season data, command area boundaries and nearness to the surface water sources.

Sources of irrigation : In the irrigated agricultural area, based on command area boundaries and nearness to the surface water sources, sources of irrigation such as tank, canal, well and irrigation were identified.

Assumptions

1. Summer image is the only criteria to decide the irrigation source (Well, Lift, canal and tank) and cropping intensity (Single or Double)

2. Lift irrigation is assumed to be along the river and some of the major canals.

3. Crops seen in the tanks and reservoirs are assumed to be as residual moisture in the soil.

4. Field knowledge is the only criteria to identify different types of plantations.

Proposed Classification :

Land Use	Season	Irrigation Status	Irrigation Source
Agriculture	Kharif	Rainfed	
		Irrigated	Well
			Tank
			Canal
			Lift
	Rabi	Rainfed	
		Irrigated	Well
			Tank
			Canal
			Lift
	Summer	Rainfed	
		Irrigated	Well
			Tank
			Canal
			Lift
	Kharif & Rabi	Rainfed	
		Irrigated	Well
			Tank
			Canal
			Lift
	Rabi & Summer	Rainfed	
		Irrigated	Well
			Tank
			Canal
			Lift
	Kharif, Rabi & Summer	Rainfed	
		Irrigated	Well
			Tank
			Canal
			Lift
Agricultural plantations			Eucalyptus
			Casuarina
			Coconut
			Mixed
Sericulture			Mulberry

Validation of the data : The generated information about each hobli was sent for validation by the respective Agricultural Officer.

Results

Land Use / Land Cover In Bangalore Rural District (2000-01)

Out of the total geographical area of 5,81,937 ha of Bangalore Rural district, area under agriculture 3,22,464 ha was (55.41%).

I.   **Area under Agriculture (Fig-1)**

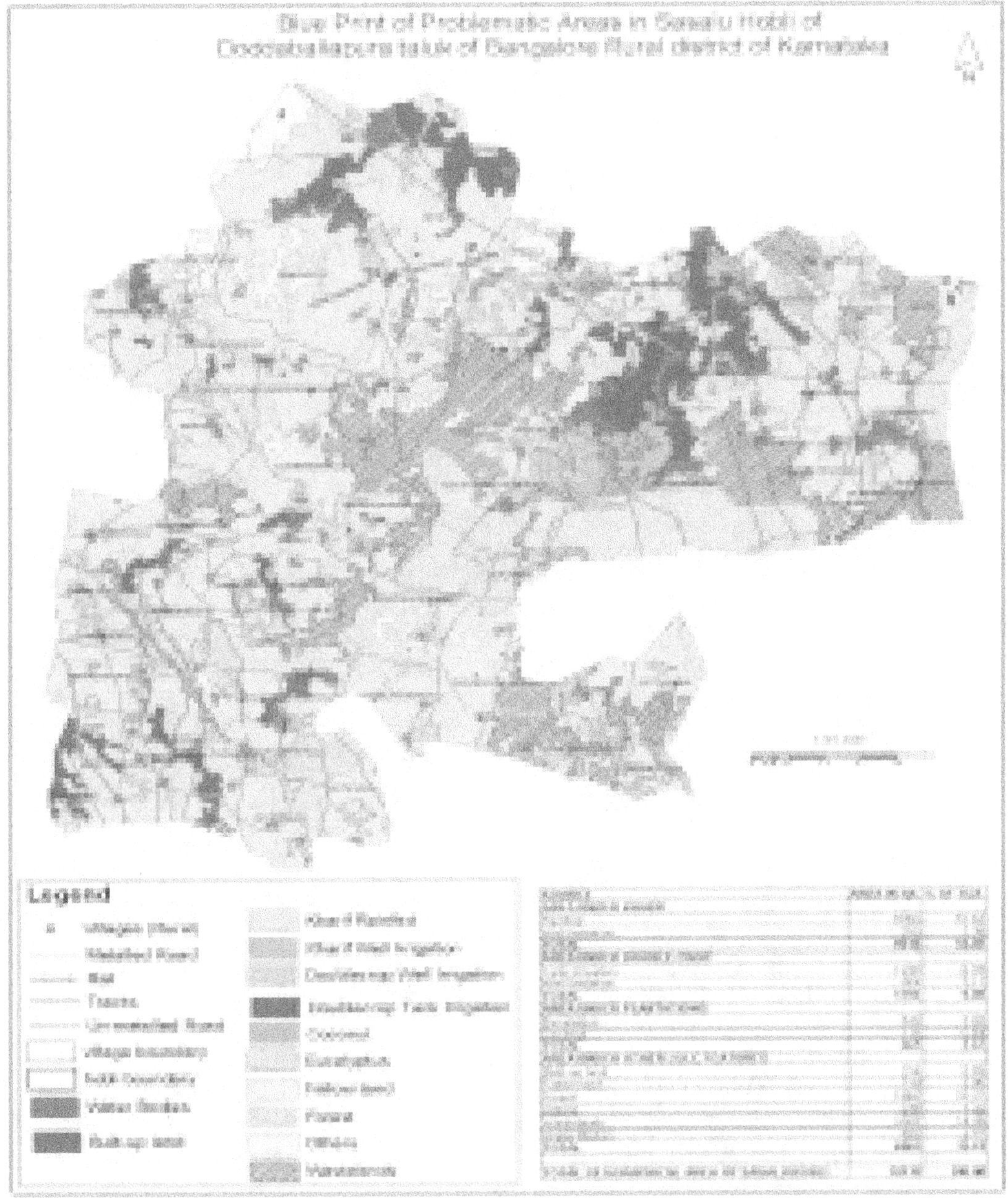

(a) Area under Kharif: The Kharif season falls in the months of June to October. Out of the total cropped area of 3,22,464 ha, 2,25,600 ha (69.96%) area was cultivated in kharif season. Kharif is an important season in Bangalore Rural district where majority of the agricultural area is cultivated under rainfed conditions. The major crops grown are ragi, rainfed paddy, jowar, maize and pulses to a minor extent. Out of the total kharif area of 2,25,000 ha, 2,21,740 ha (98.29%) was under rainfed cultivation and the remaining 1.71% (3860 ha) was under various irrigation sources such as canal, lift, well and tank (Table 1).

(b) Area under rabi: Area under exclusive rabi cultivation (November to February) was very minimal (125 ha) i.e. 0.04% of the total cropped area. Out of the rabi area of 125 ha, 117 ha was under well irrigation.

(c) Area under double crop: Double crop area includes the area where the crops were raised both during kharif and rabi reasons. In the southern part of Karnataka where red soils dominate, majority of the double cropped areas are irrigated. Out of the total cropped area of 3,22,464 ha, 96739 ha (30%) was under double crop under various sources of irrigation. The two major sources were tanks (50,204 ha, 51.90%) and wells (45,158 ha, 46.68%).

Table 1 Agricultural and use - seasons and sources of irrigation (2000-01)

SOURCE	Area (ha)	%	% to TCA
AREA UNDER KHARIF			
Rainfed	221740	98.29	68.78
Canal Irrigation	54	0.02	0.02
Well Irrigation	2291	1.02	0.71
Lift Irrigation	22	0.01	0.01
Tank Irrigation	1492	0.66	0.46
TOTAL	225600	100.00	69.96
AREA UNDER RABI			
Rainfed	7	5.93	0.00
Rabi Well Irrigation	117	94.07	0.04
TOTAL	125	100.00	0.04
AREA UNDER DOUBLE CROP			
Lift Irrigation	1258	1.30	0.39
Tank Irrigation	50204	51.90	15.57
Well Irrigation	45158	46.68	14.00
Tank and Well Irrigation	120	0.12	0.04
TOTAL	96739	100.00	30.00
TOTAL CROPPED AREA	322464		100.00

II.   Area under plantations

Plantations occupied 68,973 ha (11.85% of total geographical area) in Bangalore Rural district. The major plantations were Eucalyptus (59.56%) and Coconut (35.67%) followed by grapes, arecanut, mango and other mixed plantations. Delineation of Eucalyptus plantations area through satellite imagery helps to plan for setting up supplying agro based industries such as oil extraction and processing.

III.   Area under other Land Use / Land Cover features

Area under other land use / land cover features such as wastelands, water bodies, etc. occupied 1,90,500 ha of land (32.74% of TGA) in Bangalore Rural district. The other major features were forest (78,162 ha, 41.03%), wastelands (59,142 ha, 31.05%), water bodies (23,846 ha, 12.52%), built up land (14,825 ha, 7.78%), others (11,464 ha, 6.02%) and fallow land (3,061 ha, 1.61%) (Table 2).

Table : 2 Land use / land cover in Bangalore Rural district (2000-01)

SOURCE	Area (ha)	%	% to TGA
AREA UNDER AGRICULTURE			
Kharif	225600	69.96	38.77
Rabi	125	0.04	0.02
Double crop	96739	30.00	16.62
TOTAL	322464	100.00	55.41
AREA UNDER PLANTATIONS			
Coconut	24603	35.67	4.23
Eucalyptus	41077	59.56	7.06
Mango	89	0.13	0.02
Arecanut	720	1.04	0.12
grapes	1895	2.75	0.33
mixed	588	0.85	0.10
TOTAL	68973	100.00	11.85
AREA UNDER OTHER LULC FEATURES			
Fallow	3061	1.61	0.53
Built-up land	14825	7.78	2.55
Forest	78162	41.03	13.43
Others	11464	6.02	1.97
Wastelands	59142	31.05	10.16
Water Bodies	23846	12.52	4.10
TOTAL	190500	100.00	32.74
TOTAL GEOGRAPHICAL AREA	581937		100.00

Satellite imagery revealed that only 13.43% of total geographical area is under forest whereas per environmental norms it should be 33% of TGA. This was an alarming fact found out through remote sensing studies which stresses the impetus to be given for increasing the green cover. Population and urbanization pressure have resulted in decrease in the area of wastelands and water bodies in the district.

The information about the sources of irrigation is a crucial factor for the agricultural output. The sustenance of the well and tank irrigation depends on the rainfall. These are not assured sources unlike that under canal irrigation from reservoirs. Areas irrigated by wells and tanks and areas purely cultivated under rainfed conditions became problematic and become a cause of concern once the onset of the monsoon delays or rainfall is below the normal.

Mapping of all these layers at 1:20,000 scale has been undertaken along with statistical inputs and these maps have been given for validation to the Agricultural Officers of the 35 Raitha Samparka Kendras (Hobli level Agricultural Offices of State Department of Agriculture). All the roads, railways, tanks, canals, villages have been included in the maps so that the accuracy of the validation process increases.

An attempt has been made to compare the land use / land cover statistics of Bangalore Rural district generated by satellite imagery and by Bureau of Economics and Statistics (BES) (Tables 3 & 4). The differences observed will be checked after the maps are validated by the Agricultural Officers.

Table 3 Comparison of the land use / land cover statistics for Bangalore Rural district

LULC feature	Area (sq km)	
	Satellite data	BES
Forest	781	512
Non agriculture	386	580
Waste lands	591	720
Miscellaneous tree crops, groves	689	218
Current fallow	144	367
Net sown area	2257	2950
Total cropped area	3224	3321
Area sown more than once	967	371

Table 4 Comparison of the sources of irrigation statistics Bangalore Rural district

Source	Area (ha)	
	Satellite data	BES
Canal	54.46	9134
Tank	51696	18459
Well	47567	44496
Lift	1280	714
Others		108
Total	100597.46	72913

Conclusions

Identification of the problematic areas in agriculture in Bangalore Rural district through Remote Sensing and field validation has been undertaken to delineate the areas which need the attention of the administrators. Demarkation of the rainfed areas, areas irrigated by wells and tanks, wastelands, forests and plantations on a map of lower administrative unit such as hobli would assist the officers in several ways. The immediate advantages of such activity is assistance for planning for the locality specific and action oriented developmental activities and capacity building the of the field workers to meet the information requirements of the higher officers and the peoples representatives.

References

Anonymous, 2001, Perspective land use plan for Karnataka. State Land Use Board, Karnataka

Anonymous, 2002, Estimation of irrigated area and assessment of cropping types in selected areas of Tungabhadra Command using Remote Sensing and GIS. KSRSAC, Bangalore, Karnataka

Disaster Management

Disaster Management in India – From Space Perspective

V. Bhanumurthy, P. Manjusree

National Remote Sensing Agency, Balanagar, Hyderabad-37.
E-mail:bhanumurthy_v@nrsa.gov.in

ABSTRACT

India has been traditionally vulnerable to natural disasters like cyclone, drought, floods, earthquakes, forest fires, landslides on account its unique geographical position, climate and geological setting, Each year disasters account for loss of millions of rupees in terms of social and community assets besides economic losses that are both immediate as well as long term in nature. About 60% of our country's landmass is prone to earthquakes of varying intensities, over 40 million hectares is prone to floods, about 8% of the total area is prone to cyclones and 68% of the area is susceptible to drought. There has been an increase in the number of natural disasters over the past years, and with it, increasing losses on account of urbanization and population growth. Government of India has brought a change in the policy which emphasizes mitigation, prevention and preparedness. Our country has integrated administrative machinery for management of disasters at the national, state, district and sub-district levels. It is possible to minimize the damages with efficient disaster management plans and practices by application of modern scientific technological tools.

The most flood-prone basins in India are the Brahmaputra, Ganga and Meghana basins in the Indo-Gangetic-Brahmaputra plains in North and Northeast India, which carry 60 per cent of the nation's total river flow. Despite significant technological advances since independence, Indian agriculture continues to be periodically affected by droughts. The Indian sub-continent is subjected to varying degrees of earthquake hazard demonstrated by the fact that more than 650 earthquakes having magnitude above 5 have been recorded during the last one century. Majority of these are located in the Himalayan frontal area. About 55% of tropical deciduous forests are prone to fire. Most of the fire incidents in Indian forests are mainly man made and that too mostly initiated in grazing lands.

Space systems from their vantage position have unambiguously demonstrated their potential and capability in providing crucial information and value-added services through remote sensing and communication technology for disaster management. The earth observation satellites provide a comprehensive, synoptic and multi-temporal coverage of large areas in real time and at frequent intervals providing continuous monitoring of atmospheric as well surface parameters related to natural disasters. Geo-stationary satellites provide continuous and synoptic observations over large areas on weather including cyclone monitoring, Polar orbiting satellites have the advantage of providing much higher resolution images, even though at low temporal frequency, which could be used for, detailed monitoring, damage assessment and long-term relief management. Communication and networking plays a very prominent role in the crisis of a disaster. The advantage of satellite remote sensing is that the data are reliable, scientific, accurate, fast and economic. Any disaster cycle has four phases namely, Forecasting & Warning, Relief, Rescue & Rehabilitation and Mitigation. The role of satellite remote sensing data in different phases of disaster cycle is discussed here in this paper along with a few case studies.

Introduction

India, on account of its geographical position, climate and geological setting, is the worst affected centre of disasters in the South Asian region, making it vulnerable to natural hazards such as cyclone, drought, floods, earthquakes, forest fires, landslides and avalanches. Each year disasters account for loss of millions of rupees in terms of social and community assets. Disaster management occupies an important place in the country's policy framework as it is the poor and the under-privileged who are worst affected on account of calamities/disasters. Therefore, Government of India has brought a change in the policy which emphasizes mitigation, prevention and preparedness. Our country has integrated administrative machinery for management of disasters at the National, State, District and Sub-district levels.

About 60% of the country's landmass is in the seismic zones III-IV and prone to earthquakes of varying intensities, sub-Himalayan/western ghat is vulnerable to landslides, over 40 million hectares is prone to floods, about 8% of the total area, in particular the Eastern Coastal states are prone to cyclones and 68% of the net sown area is susceptible to drought. The annual average cropped area affected due to floods is approximately 3.7 million hectares. The most flood-prone basins in India are the Brahmaputra, Ganga and Meghana basins in the Indo-Gangetic-Brahmaputra plains in North and Northeast India, which carry 60 per cent of the nation's total river flow. Since nearly 75 per cent of the total rainfall is concentrated over a short monsoon season of four months (June-September), the rivers witness a heavy discharge during these months, leading to widespread floods. The rivers originating in the Himalayas cause erosion of the banks in the upper reaches and over-topping in the lower reaches.

The colossal damage caused by the above cited natural disasters, though, cannot be nullified, yet they can be mitigated by integrating new and existing technology and expertise, and by managing risk through various structural and non-structural strategies. Flood disaster management demands efficient planning measures, implementation and policy making decisions. Application of modern scientific and communication tools in the existing management practices will help in smooth functioning of the system. One of the most important elements in disaster management is the availability of timely information for taking decisions and actions. Hence, the necessity of a well-defined system has become obvious to address the various information needs and to provide an operational service with its framework

Space technology

Space technology in the form of remote sensing and communication satellites are being extensively used for managing disasters in the recent years. The advantage of satellite remote sensing is that the data is reliable, scientific, accurate, fast and economic. Communication satellites are widely used for timely dissemination of early warning and real-time co-ordination of relief operations. The operational role of satellite communications in providing the emergency communications viz., satellite phones, point-to-point networking solutions routed through the arrays of VSATs deployments in remote and inaccessible areas, Cyclone Warning and Dissemination Systems (CWDS), Data Collection Platforms (DCP) and Satellite Aided Search & Rescue (SAS&R) are critical if there are well-knit institutional mechanisms to integrate the operational services emanating from these technologies for operational purposes.

Role of space applications in disaster management lies in its criticality to produce as well as disseminate the information on real/near real time basis. In last couple of decades, space applications to disaster management have gone through the phases of experimental demonstration, semi-operational and operationalisation in certain areas of disaster management. However, the real strength of theses applications could be realized only in the synergy and convergence with other collateral information as well as in tune with traditional technologies. The role of space applications related agencies assumes vitality and holds significance in terms of enabling this to happen and aiding the appropriate value so that the final delivered products could be in harmony with the operational needs of the disaster management community down the line and are of operational importance.

Geographic Information system (GIS) is a powerful tool in which spatial information can be stored, organized, and retrieved in a user friendly format. Conjunction of satellite remote sensing data and ancillary data in GIS environment combined with the Global positioning system (GPS) data is a potential tool to a disaster manager. The awareness and utilization of these technologies and the power of spatial information systems specifically oriented towards decision making or resource management is growing rapidly in our country. In this direction, a web-enabled GIS-based resource inventory called India Disaster Resource Network (IDRN) was set up by Ministry of Home Affairs, Government of India, which lists out the necessary resources for emergency response available at the district and state level through out the country so that resources can be mobilized at a short notice.

Disaster Cycle

The developments in space technology offer tremendous technological potential to address the critical information needs during all the phases of disaster management, which include mitigation and preparedness, response and recovery/relief. Any disaster cycle consists of four phases namely, Forecasting & Warning, Relief & Rescue, Rehabilitation and Mitigation (FWR^{3}M). Satellite remote sensing plays a potential role in the disaster cycle by providing information in almost all the phases of disaster. Fig-1 shows the schematic view of a disaster cycle.

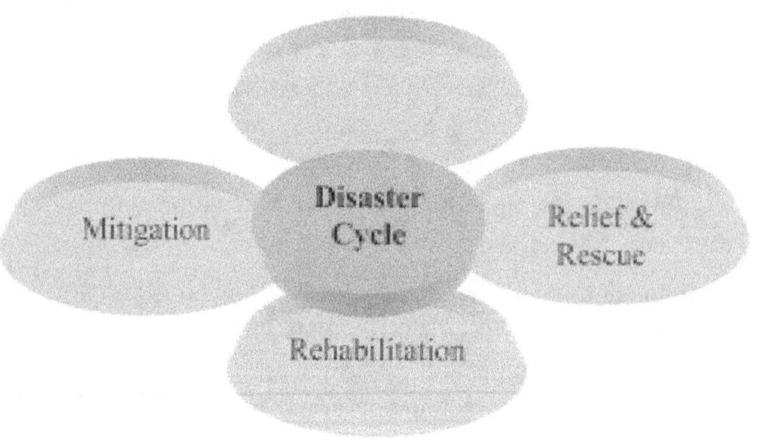

Fig. 1 Schematic view of a disaster cycle

Forecasting and Warning

Forecasting and warning is the most important phase in a disaster cycle wherein the occurrence of a disaster is foreseen before it occurs using the models and the information is disseminated to the public people for evacuation. In our country, Central Water Commission, a nodal agency under Government of India is responsible for issuing flood forecasting bulletins on different river basins. Rainfall-runoff and flood routing models incorporate hydrological and meteorological information about the basin including the landuse/landcover, soil texture, slope of the terrain, drainage network, rainfall etc. All these parameters can be obtained from satellite data. From satellite data like METEOSAT, DMSP, TRMM, GPM etc, it is possible to obtain spatial distribution of rainfall, which can be incorporated in the models. Information on land use / land cover, soil, drainage is available from the medium-resolution satellites and Digital Elevation Model (DEM) from satellites like SRTM, ASTER etc respectively.

Satellite communication provides an effective mechanism for real time dissemination of information and early warning besides establishing communication link after a disaster like cyclone. Indian Meteorological Department (IMD), Government of India is responsible for issuing the cyclone warnings in our country. The geo- stationary satellites like INSAT system provide half hourly observation of cyclone development and movement and its associated parameter for warning and prediction of landfall. They are more useful particularly when the tropical cyclones are beyond the range of the coastal radars. The Cyclone Warning Dissemination System (CWDS) which are situated along the Indian Eastern coastal line disseminates warnings of an impending event to village administration, District collectors, State Govt. officials etc.

The conventional methods for monitoring of both causative factors as well as impact of agricultural drought suffer from various limitations such as sparse observations, subjective data etc. Satellite sensors provide direct spatial information on vegetation stress caused by drought conditions and the seasonal progression of Normalised Difference Vegetation Index (NDVI) compared to normal NDVI profile are utilised in the assessment of agricultural drought using the ground data on rainfall, crop sowing progress and reservoir levels in a complementary manner. The relative deviation of NDVI from that of normal and the rate of progression of NDVI during the season give the indication about the agricultural situation in the district which is then complemented by ground situation as evident from rainfall and sown area.

Using satellite data, the extent of forest burnt-areas and damage assessment can be carried out. The forest fire occurs over several places throughout the fire season (January to May), the period of monitoring is necessary to understand the areas burnt during different months, peak month of burning, recovery levels etc. Such information would be useful in planning effective control operations and damage assessment. Continuous fire burnt areas monitoring using high temporal satellite data like IRS P6/1D AWiFS/ WiFS helps in assessment of progression of fires at frequent intervals (1-3 days) at forest range level to facilitate control operations and damage assessment.

Seismological observations in the country are made through national network of 36 seismic stations operated by the IMD, which is the nodal agency. These stations have collected data over long periods of time.

Relief and Rescue

Relief and rescue phase activities resume during and immediately after the disaster. They are designed to provide emergency help to the affected people and reduce the likelihood of secondary damage. Immediately after a flood occurs, the top most priority information required is the spatial extent of the affected area on the ground. This information about the flood situation on the ground needs to be continuously updated for the successful execution of the operations. This information is provided by a flood map prepared from satellite remote sensing in a synoptic and a cost effective manner. The flood inundation information can be provided at regular intervals of time to the concerned departments.

With the combination of satellites onboard, including optical and microwave, it is possible to generate flood inundation information temporally and accurately in a scientific way. Indian Remote Sensing satellite (IRS) series have WiFS / AWiFS sensor that have a coarse spatial resolution of 188/58m, a large swath and a high repetitivity. They also have PAN and MX camera that has a fine resolution of 5.8 m in mono and multi-spectral mode. This facilitates monitoring of the flood affected area state wise at macro level and detailed study of worst affected areas at micro level.

Satellite derived NDVI along with other parameters such as crop condition, together with ground observations will be a valuable input for assessing damages/yield for crops in case of drought. For landslides and earthquakes, emergency communication support and damage assessment through aerial surveys over the worst affected areas is the integral part that a disaster manager requires. Toward this, satellite remote sensing plays an important role in the mapping of active faults, lineament and density of lineaments. The availability of very high-resolution data can be made use for post-disaster assessment in the dense urban clusters. This data gives a synoptic overview of the area affected by the disaster. Such data can be made use to create a very large-scale base information of the terrain for carrying out the disaster assessment and for relief measures. It can also map some geomorphic changes, if any, in the micro-level after the earthquake for micro-seismic zonation.

Rehabilitation and Mitigation

In the rehabilitation phase, information on the uplands for construction of shelters is very important for a disaster manager. High resolution satellite data can be used to generate information on the landuse/landcover including uplands for the complete basin. Identification of safer locations/ reconstruction of shelters for rehabilitation can be carried out using satellite data and ortho-photos from aerial data for landslides, earthquakes and cyclones. Mitigation measures are executed by the flood control departments to eliminate or reduce the chance of occurrence of a flood which can be classified into structural and non-structural methods. Some of the structural methods are construction of embankments (dikes or levees) and floodwalls confined to the floodwaters, the improvement of river channels by straightening/ widening /deepening construction of bypass and diversion channels and construction of reservoirs for the temporary storage of floodwaters. Some of the non-structural methods are flood forecasting, flood plain zoning, flood insurance, post-flood rescue and rehabilitation measures. Further regular monitoring of the vulnerability of existing flood protective control structures, planning of future control structures like embankments and spurs which depend on the behavior of the river, river bank protection measures are carried out annually.

Spatial information on the flood affected area and special characteristic of each river are vital for planning of flood protective control structures. Temporal and historic high resolution data provides details on the changes in the river channels over the years and river bank erosion and deposition. The vulnerable pockets of erosion can also be predicted using the satellite data so as to plan suitable river bank protection measures. Using historic satellite data combined with hydrological data, a flood hazard zone map can be prepared for each flood prone basin. These maps are useful to make policy decisions and regulate action plans. A flood hazard map can be integrated with the economic value of the flood plain and a flood risk map can be prepared which helps in providing the flood insurance.

For drought, space based inputs can be used in the long-term planning of drought combating measures like soil/water conservation, check dams, water harvesting structures, ground water development.

For generating landslide hazard zonation maps, the critical terrain information like the updated lithology, geological structure, geomorphology, land use / land cover and drainage derived from the satellite data are integrated with other topographic data like the slope, aspect and morphology. Where failure could occur can be addressed in a more regional geographic information system (GIS) analysis as a necessary first step in risk analysis. These can include an inventory of landslides; seismic records; large-scale geological mapping, extensive geotechnical data on rock properties; high-resolution digital elevation data, suitable high-resolution remote sensing data and aerial photographs. Detailed slope information is essential for reliable landslide inventory maps. Currently, topographic maps and digital elevation data are used. Slope also effects surface drainage and is an important factor in the stability of the land surface. Current research has shown that airborne and satellite InSAR techniques are being used to produce detailed slope information. This allows a more accurate interpretation of slope morphology and regional fracture systems with topographic expressions. Based on the landslide hazard zonation images along with the local condition and terrain features, management maps can be prepared for reducing the risk.

Experience of ISRO/DOS
In order to provide vital inputs and support in the event of disasters, ISRO/DOS has been pursuing the development of techniques and methodology - integrating space based systems and services for disaster management. Over the past two decades, DOS has carried out several studies in providing space-based inputs to disaster management. Satellite based mapping of all the major floods and damage assessment is being carried out by National Remote Sensing Agency, (NRSA) Hyderabad. This crucial flood inundation information is disseminated to central and concerned state governments by digital means. The data from Indian Remote Sensing satellites and RADARSAT satellite (microwave) is used to map the

flood-inundated areas in near real-time and estimate the damage to crops. All major cyclones Orissa Super Cyclone in October 1999, Gujarat (Kandla) cyclone in 1998, Andhra Pradesh cyclone in 1996 etc that have occurred during the last one decade have been monitored. During the Orissa Super Cyclone in October 1999, space inputs/services were provided in terms of establishing a V-SAT based emergency communication network and remote sensing based estimation of inundated areas and crop damage.

National Agricultural Drought Assessment and Monitoring System (NADAMS) is an operational remote sensing based agricultural drought monitoring mechanism in India at NRSA and provide near real-time information on prevalence, severity level and persistence of agricultural drought at national/state/district level. The project covers 14 states of India which are predominantly agriculture based and prone to drought situation. Satellite sensors provide direct spatial information on vegetation stress caused by drought conditions and the seasonal progression of NDVI compared to normal NDVI profile are utilised in the assessment of agricultural drought using the ground data on rainfall, crop sowing progress and reservoir levels in a complementary manner. Periodic monthly/fortnightly bulletins are being issued to the officials of the line departments of the country.

NRSA/DOS has been regularly mapping major forest fire burnt area assessment in the country and monitoring their progression using high resolution satellite data like IRS LISS III/LISS IV. This kind of information is necessary in the event of episodic large area forest fires, for timely proper protection of biodiversity rich areas, high regeneration areas and also wildlife protection planning. Earlier studies conducted by NRSA revealed the feasibility of establishing fire frequent zones using the last 10-15 years historical data. Information derived from satellite data is one of the vital inputs for installation of fire watch towers based on fire visibility analysis in the DEM environment.

The main contribution of satellite data on landslides is to provide the geological details, morphological, and land use, in determining how the landslide occurs and what causes the failure. A project on mapping of the potential landslide hazard zones in parts of Uttaranchal and Himachal Pradesh (along the pilgrimage routes) was undertaken and completed. For selected areas landlside hazard zonation maps were prepared.

Remote sensing data can provide the basic inputs on the structural fabric of the terrain. The lineament map is one of the important inputs for delineating the seismo-tectonic province. As observed in the study of Bhuj earthquake, Chamoli earthquake and Latur, most of the epicenters are located nearer to such structurally weak zones. The area affected by earthquakes are generally large, but they are restricted to well known regions near plate contacts. The associated surface manifestation of such disaster is fault rupture damage due to ground shaking, liquefaction and landslides. Remote sensing provides vital inputs regarding the geology and topographic conditions, which can be used as inputs for seismic hazard zonation. In association with the collateral data, information on contrasting lithology, faults and lineaments, geomorphic contrast, demarcation of soft sediments, surface deformation, etc. can be generated, which is useful in better management of earthquake hazards.

Case Studies

Near Real Time Flood Monitoring and Mapping

For the past one and half decades, National Remote Sensing Agency, Dept. of Space, Hyderabad, is carrying out this activity, using satellite remote sensing data. Major floods in the country are mapped, monitored and flood maps showing the spatial flood extent on the ground are prepared. Fig-2 shows major flood prone states in the country. Flood damage information like the flood inundated area, submerged crop area number of villages marooned and affected transport network is generated by application of appropriate models,. The flood maps along with the flood damage information are sent to the departments concerned like Ministry of Home Affairs, Central Water Commission, Central and State relief departments and Flood Control departments and even to district collectors at times.

Fig. 2 Major flood-prone states in the country

The activity comprises of four major components namely the flood watch, satellite data planning and acquisition, satellite data analysis and output data dissemination to the users. Fig-3 shows the illustration of the components of this activity.

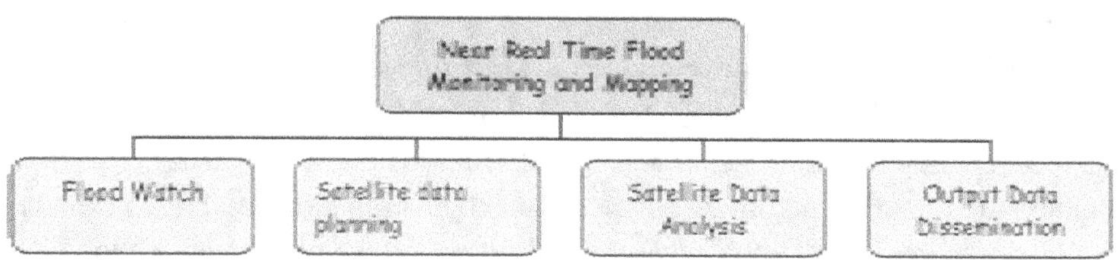

Fig. 3 Components of the Near Real Time Flood Monitoring and Mapping Activity

The flood situation in the country is monitored through news media, line departments, Internet and ground hydrological and meteorological data. The trend of the flood wave is analysed from the river gauge hydrographs which are prepared from the daily water level data for the major flood prone rivers like Brahmaputra and Ganga. The daily rainfall data is also checked for any heavy rainfall anomalies. Satellite data planning is a very crucial component which demands meticulous planning for programming and tilting of satellite sensors on the affected area. A flood disaster watch report is prepared showing the details of the occurrence of the flood and the satellite data acquisition plans. All the procured satellite data is analysed and flood inundation layer is extracted. A flood map is composed by overlaying this layer on a topo map along with the pre-flood information. The single-bit extracted flood inundation layer is integrated with the database to generate district/village-wise flood inundated area statistics, submerged crop area/villages. The turn-around-time achieved in generating this information is less than 5 hours. The flood maps and the flood inundation information is disseminated to the user departments immediately after generation through electronic means to Ministry of Home Affairs, Central Water Commission, Central and state relief commissioners, besides other user departments.

Bihar Floods-2004

Floods were reported during last week of June, 2004 in Bihar due to continuous rains. Over two crore people were reported to be affected due to the floods. According to the Central Water Commission, the rivers of Burhi Gandak, Bagmati, Adhwara, Gandak, Kamla Balan and Mahananda flowed above the danger mark all along their course. The overall flood situation in the flood affected districts improved by the first week of August, 2004, with major rivers and their tributaries receding. Around 22 districts were affected by the flood and the flood-hit districts were Sheohar, Sitamarhi Supaul, Saharsa, East Champaran, West Champaran, Vaishali, Samastipur, Khagaria, Araria, Kishanganj, Bhagalpur, Saran, Gopalganj, Begusarai, Siwan, Muzaffarpur, Darbhanga, Katihar , Madhubani, Purnia and Madhepura.

Flood Disaster Team of Decision Support Centre, National Remote Sensing Agency, Dept. of Space, Govt. of India kept a constant watch on the flood situation and checked the satellite data coverage over the flood affected areas. Satellite data both optical and microwave from Indian remote sensing satellite IRS-P4 OCM and Canadian satellite- Radarsat was acquired over flood affected areas respectively. IRS-P6 AWiFS data of 20-Feb-2004 was procured to delineate the pre-flood river bank and active river channel (Fig4a). The microwave

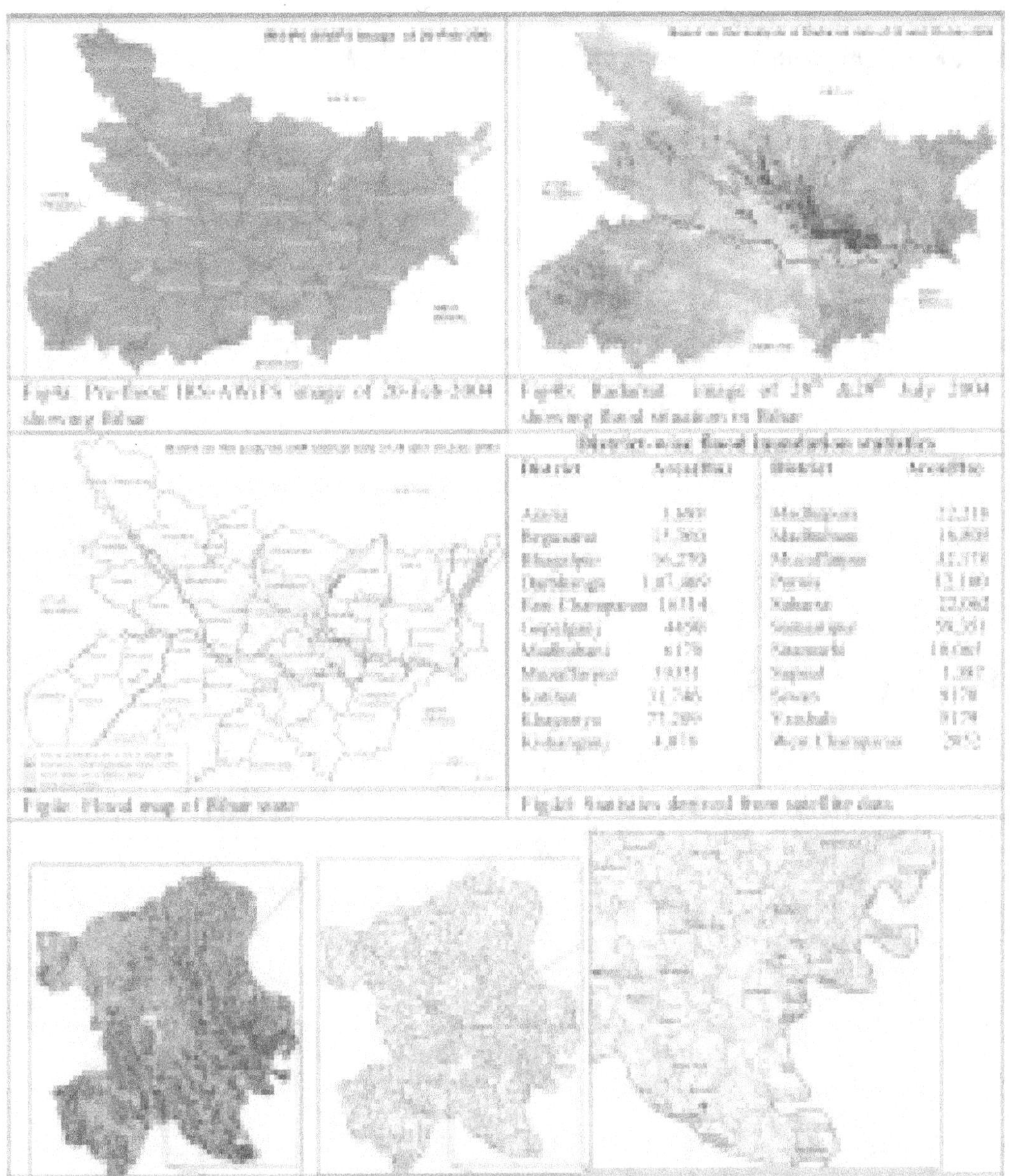

Fig. 4e Flooded villages in Sitamarhi district as on 14th *July 2004*

data of 28th and 29th July 2004 from Radarsat satellite was combined (Fig4b) and analysed. Rapid flood damage assessment was carried out and flood map as well as district-wise flood inundation statistics were sent to the concerned users on the same day of acquisition (Fig4c&d). Detailed damage assessment was also carried out and maps showing villages marooned were prepared (Fig4e).

5.2 Landslide Hazard Zonation

A project on mapping of the potential landslide hazard zones in parts of Uttaranchal and Himachal Pradesh (along the pilgrimage routes) was undertaken. For generating landslide hazard zonation maps, the critical terrain information like the updated lithology, geological structure, geomorphology, land use / land cover and drainage derived from the satellite data was integrated with other topographic data like the slope, aspect and morphology using GIS models. These hazard risk maps will assist in emergency preparedness, planning and in making rational decisions regarding development and construction in areas susceptible to slope failure. Fig5a shows the landslide hazard zonation for Uttar Kashi area and Fig 5b shows the landslide affected zone near Vishnuprayag on July 06, 2004.

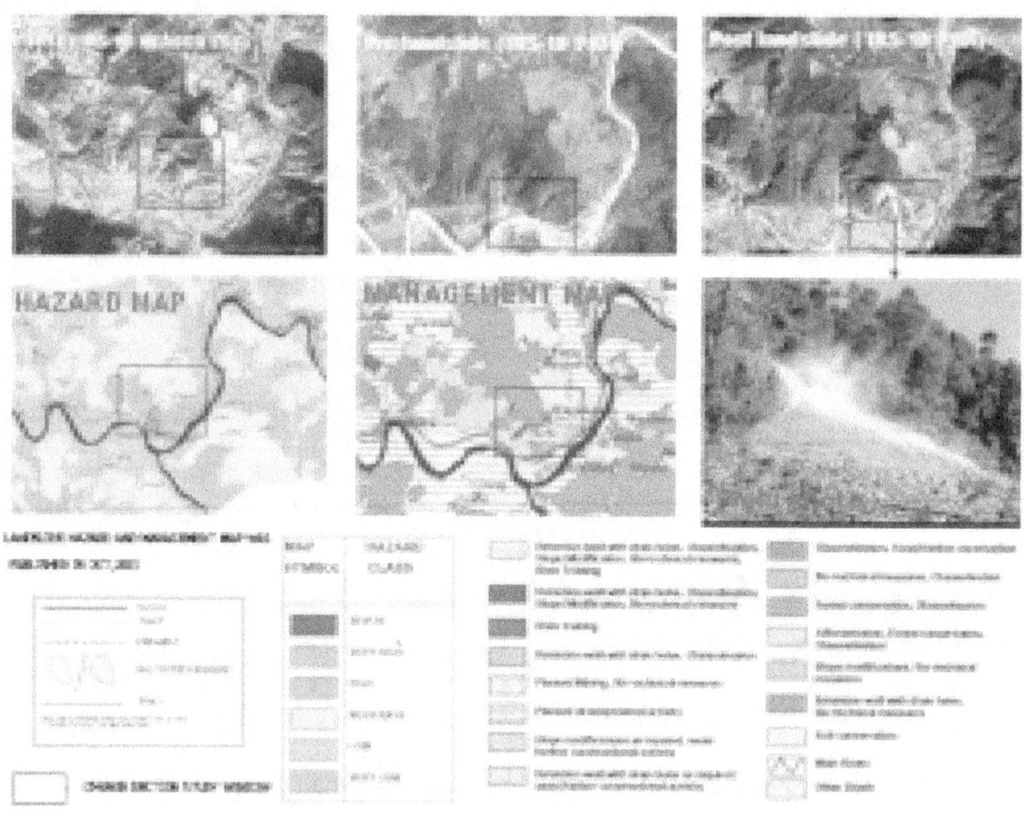

Fig. 5a Landslide hazard zonation for Uttar Kashi area

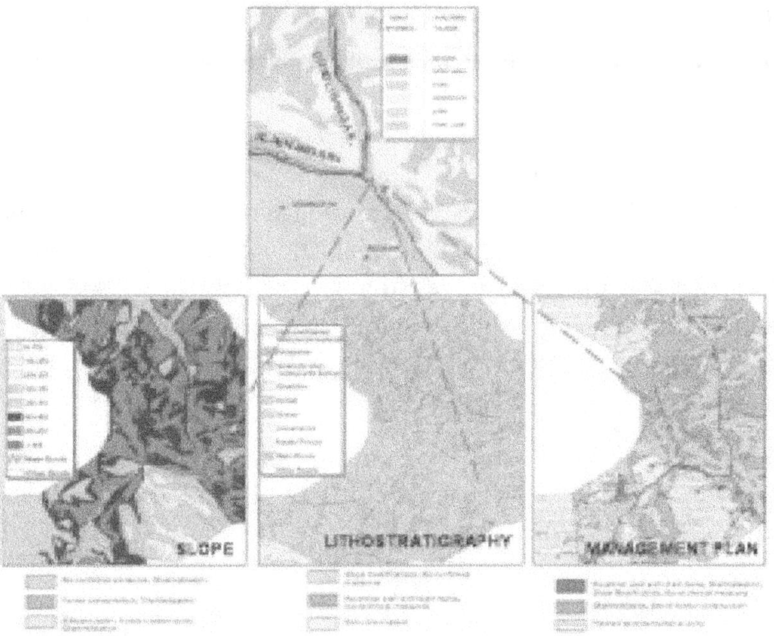

Fig. 5b Landslide affected zone near Vishnuprayag on July 06, 2004

Drought Assessment

National Agricultural Drought Assessment and Monitoring System (NADAMS) is an operational remote sensing based agricultural drought monitoring mechanism in India at NRSA and provide near real-time information on prevalence, severity level and persistence of agricultural drought at national/state/district level. The project covers 14 states of India which are predominantly agriculture based and prone to drought situation. The assessment of agricultural drought situation depends on (1) seasonal NDVI progression – i.e. transformation of NDVI from the beginning of the season over agricultural areas, (2) comparison of agricultural area NDVI profile with previous normal years, (3) weekly rainfall status compared to normal, and (4) weekly progression of sown area compared to normal. Periodic monthly/fortnightly bulletins are being issued to the officials of the line departments of the country.

The vegetation index image which represents the vegetation status will be derived from the AVHRR sensor (Advanced Very High Resolution Radiometer) which is aboard NOAA (National Oceanic and Atmospheric Administration) satellite. Normalized Difference Vegetation Index (NDVI) is derived from the equation (bnd2-bnd1)/ (bnd2+bnd1), where bnd1 and bnd2 are the reflected radiation in visible and near infrared channels of NOAA AVHRR sensor. The legend of colour bars and the vegetation index values are provided along with the vegetation index image. Colours from yellow through Green to Red indicate the increasing vegetation vigor. The agriculture crop condition is monitored based on NDVI information derived from satellite data. After precluding the forest area, district-wise average vegetation index statistics were generated.

During 2004, 14 states were affected with agriculture drought namely Andhra Pradesh, Bihar, Chattisgarh, Gujarat, Haryana, Jharkhand, Karnataka, Maharashtra, M.P, Orissa, Rajasthan, Tamil Nadu, Uttaranchal and Uttar Pradesh and were mapped for the period June to November 2004. Country, state and district-level assessment and reporting was made on monthly basis. NDVI analysis addresses about 40% of the drought condition. The salient features of 2004 drought assessment are that normal trend of agricultural vegetation development indicating normal agricultural situation in most parts of the country. Poor vegetation status was observed in few districts of Rajasthan state. IRS WiFS based assessment in Andhra Pradesh and Karnataka states indicate normal agricultural situation in most parts. However in some of the districts of Andhra Pradesh delay in seasonal vegetation development was evident.

Fig 6a shows the vegetation index image of the country derived from NOAA AVHRR satellite data analysis during 2004 and Fig 6b shows the vegetation index image of Guntur district derived from IRS- WiFS data.

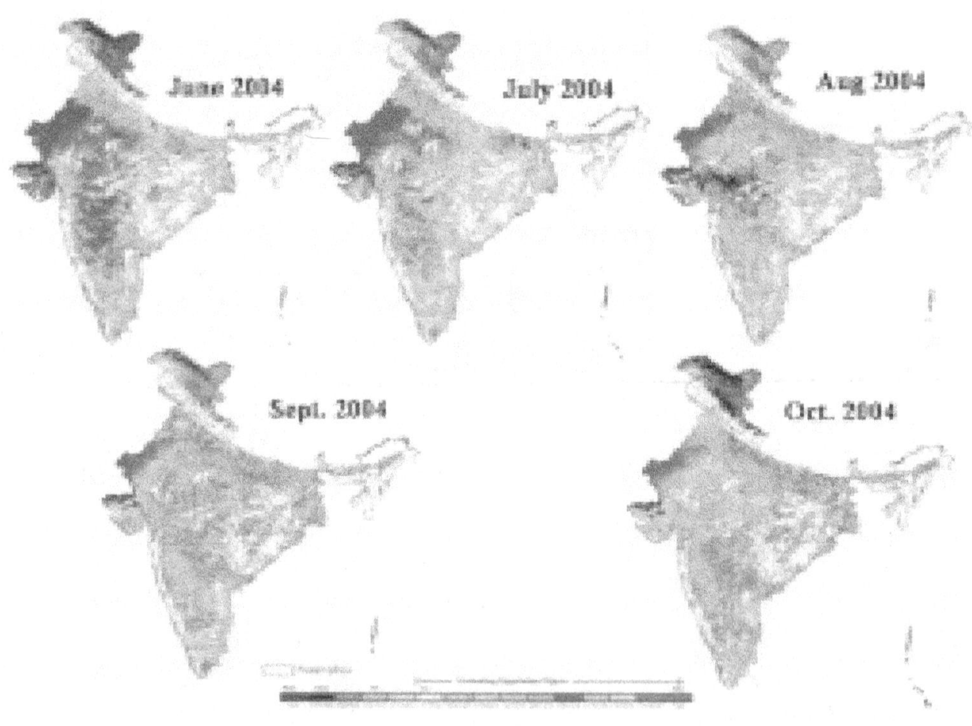

Fig. 6a NDVI image of the country during 2004

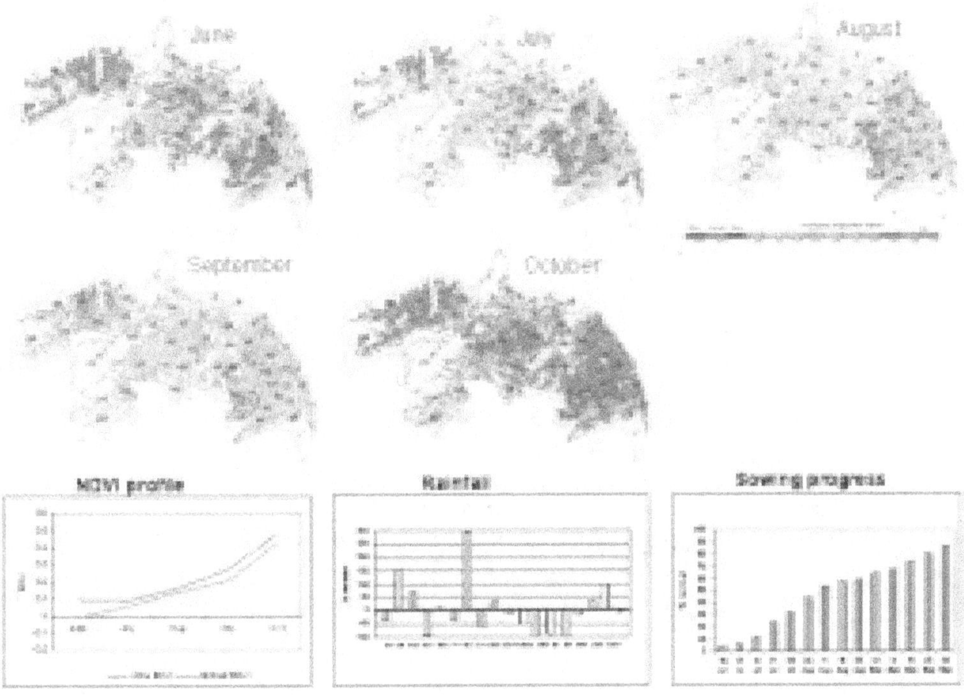

Fig. 6b NDVI image of Guntur district in Andhra Pradesh during Kharif 2004

Conclusion

Space technology has its own potential role in the relief, rehabilitation, mitigation and forecasting phases of disaster management as demonstrated here in this paper. Application of the latest technologies in the conventional practices will help in reducing the impact of disasters to some extent. Research activities are being taken by various academic institutions across the country on inundation simulation using flood stage & discharge data and close contour information and generation of Digital Flood Insurance Rate Maps. Research is also being carried out on rainfall estimation from satellite data to supplement ground observations, development of procedures for early drought warning integrating data from multiple sources, automated forest fire detection and progression, spatial modelling of forest fires and GIS based data integration, precursor studies on before earthquake (thermal and ionospheric disturbances), SAR interferometry and Geodetic GPS for monitoring active faults, peak ground acceleration and seismic hazard zonation, With more advances in the space technology stored in future, with sophisticated sensors and more capabilities, it is possible to make a safer India from natural disasters.

References

1. Disaster Management in India, Publication of Ministry of Home affairs, Govt. of India

2. NADAMS monthly drought reports of Kharif 2004.

3. C.S.Murthy.,et al., Concepts and overview of NADAMS, UNESCAP/NRSA regional workshop on Agricultural drought monitoring & Assessment using space technology, May3-7, 2004, Hyd.

4. Rao.D.P., 2000, Disaster Management, GIS development, March 2000, Vol.IV, Issue3

5. Vinod kumar.K, T.R.Martha, Evaluation of IRS-P6 data for geological studies, NNRMS Bulletin, Sept 2004

6. Rao.D.P., 2000, Remote Sensing and GIS for sustainable development, Geospatial today, May-June 2002, Vol.1, Issue1

7. Myint, A. K. and Hofer, T., 1997, Forestry and Key Asian Watersheds, ICIMOD, Kathmandu

8. Rao, D.P., 2000, Disaster Management, Proceedings of Map India 2000. April 10-11, 2000, New Delhi, pp. NDM-5

9. Rao, M.S., 1989, Keynote Address on Floods, Symposium on Preparedness, Mitigation and Management of Natural Disasters, New Delhi, August 2-4

10. *Wolman, M. G., Factors influencing erosion of a cohesive riverbank.* Am. J. Sci., *1959, 257, 204.*

11. Agricultural drought assessment using remotely sensed data: Indian scenario, Second conference on Disaster management case histories held at Pilani during November 14-16 2003.